The Great Gray Plague

by

Raymond F. Jones

Double 9 BOOKS

The Great Gray Plague

by Raymond F. Jones

ISBN: 978-93-59324-24-1

Published by

DOUBLE 9 BOOKS
2/13-B, Ansari Road, Daryaganj
New Delhi - 110002
info@double9books.com
www.double9books.com
Tel. 011-40042856

ABOUT THE AUTHOR

Raymond Fisher Jones (15 November 1915 - 24 January 1994) was a science fiction author from the United States. He is most known for his 1952 novel This Island Earth, which was made into the 1955 film of the same name. Jones was born in Salt Lake City, Utah, and has always been a member of the Church of Jesus Christ of Latter-day Saints. In 1994, he died in Sandy, Utah. The majority of Jones' short fiction was published in publications such as Thrilling Wonder Stories, Astounding Stories, and Galaxy during the 1940s, 1950s, and 1960s. Between 1951 and 1978, he authored sixteen novels. His short tale "Rat Race," which initially appeared in the April 1966 edition of Analog Science Fiction and Fact, received a Hugo Award nomination. "Correspondence Course," which first appeared in the April 1945 issue of Astounding Stories, was nominated for a Retro Hugo award for best short story in 1996. Another short story, "The Alien Machine," which first appeared in the June 1949 issue of Thrilling Wonder Stories, was later combined with two other short stories, "The Shroud of Secrecy" and "The Greater Conflict," and expanded into the novel This Island Earth, which inspired the film of the same name.

Dr. William Baker was fifty and didn't mind it a bit. Fifty was a tremendously satisfying age. With that exact number of years behind him a man had stature that could be had in no other way. Younger men, who achieve vast things at, say, thirty-five, are always spoken of with their age as a factor. And no matter what the intent of the connection, when a man's accomplishments are linked to the number of years since he was born there is always a sense of apologia about it.

But when a man is fifty his age is no longer mentioned. His name stands alone on whatever foundation his achievements have provided. He has stature without apology, if the years have been profitably spent.

William Baker considered his years had been very profitably spent. He had achieved the Ph. D. and the D. Sc. degrees in the widely separated fields of electronics and chemistry. He had been responsible for some of the most important radar developments of the World War II period. And now he held a post that was the crowning achievement of those years of study and effort.

On this day of his fiftieth birthday he walked briskly along the corridor of the Bureau building. He paused only when he came to the glass door which was lettered in gold: National Bureau of Scientific Development, Dr. William Baker, Director. He was unable to regard that door without a sense of pride. But he was convinced the pride was thoroughly justifiable.

He turned the knob and stepped into the office. Then his brisk stride came to a pause. He closed the door slowly and frowned. The room was empty. Neither his receptionist nor his secretary, who should have been visible in the adjoining room, were at their posts. Through the other open door, at his left, he could see that his administrative assistant, Dr. James Pehrson, was not at his desk.

He had always expected his staff to be punctual. In annoyance that took some of the glint off this day, he twisted the knob of his own office door and strode in.

He stopped just inside the room, and a warm wave of affection welled up within him. All nine members of his immediate staff were gathered around the table in the center of his office. On the table was a cake with pink frosting. A single golden candle burned brightly in the middle of the inscription: Happy Birthday, Chief.

The staff broke into a frighteningly off-key rendition of "Happy Birthday to You." William Baker smiled fondly, catching the eye of each of them as they badgered the song to its conclusion.

Afterward, he stood for a moment, aware of the moisture in his own eyes, then said quietly, "Thank you. Thank you very much, Family. This is most unexpected. None of you will ever know how much I appreciate your thoughtfulness."

"Don't go away," said Doris Quist, his blond and efficient secretary. "There's more. This is from all of us."

He opened the package she offered him. A genuine leather brief case. Of course, the Government didn't approve of gifts like this. If he observed the rules strictly, he ought to decline the gift, but he just couldn't do that. The faces of Doris and the others were glowing as he held up the magnificent brief case. This was the first time such a thing had occurred in his office, and a man hit fifty only once.

"Thanks so much for remembering," Baker said. "Things like this and people like you make it all worth while."

When they were all gone he sat down at his desk to take up the day's routine. He felt a little twinge of guilt at the great satisfaction that filled him. But he couldn't help it. A fine family, an excellent professional position—a position of prominence and authority in the field that interested him most—what more could a man want?

His meditation was interrupted by the buzzing of the interphone. Pehrson was on the other end. "Just reminding you, Chief," the assistant said. "Dr. Fenwick will be in at nine-thirty regarding the request for the Clearwater grant. Would you like to review the file before he arrives?"

"Yes, please," said Baker. "Bring everything in. There's been no change, no new information, I suppose?"

"I'm afraid not. The Index is hopelessly low. In view of that fact there can be no answer but a negative one. I'm sorry."

"It's all right. I can make Fenwick understand, I'm sure. It may take a little time, and he may erupt a bit, but it'll work out."

Baker cut off and waited while Pehrson came in silently and laid the file folders of the offending case on the desk. Pehrson was the epitome of owl-eyed efficiency, but now he showed sympathy behind his great horn-rimmed spectacles as he considered Baker's plight. "I wish we could find some way to make the Clearwater research grant," he said. "With just a couple of good Ph. D.'s who had published a few things, the Index would be high enough—"

"It doesn't matter. Fenwick is capable of handling his own troubles." Pehrson was a good man, but this kind of solicitousness Baker found annoying.

"I'll send him in as soon as he comes," Pehrson said as he closed the door behind him.

Baker sighed as he glanced at the folder labeled, Clearwater College. Jerkwater is what it should be, he thought. He almost wished he had let Pehrson handle Fenwick. But one couldn't neglect old friends, even though there was nothing that could be done for shortsighted ones.

Baker's memories shifted. He and Fenwick had gone to school together. Fenwick had always been one to get off into weird wide alleys, mostly dead ended. Now he was involved in what was probably the most dead ended of all. For the last three years he had been president of little Jerkwater—Clearwater College, and he seemed to have some hope that NBSD could help him out of the hole.

That was a mistake many people made. Baker sometimes felt that half his time was spent in explaining that NBSD was not in the business of helping people and institutions out of holes. It was in the business of buying for the United States Government the best scientific research available in the world.

Fenwick wanted help that would put Clearwater College on its feet through a research contract in solid state physics. Fenwick, thought Baker, was dreaming. But that was Fenwick.

The President of Clearwater College entered the outer office promptly at nine-thirty. Pehrson greeted him, and Doris showed him into Baker's office.

Dr. John Fenwick didn't look like a college president, and Baker, unknowingly, held this vaguely against him, too. He looked more like a prosperous small business man and gave the impression of having just finished a brisk workout on the handball court, and a cold shower. He was ruddy and robust and ill-equipped with academic dignity.

Baker pumped his hand as if genuinely glad to see him. "It's good to see you again, John. Come on over and sit down."

"I'll bet you hoped I'd break a leg on the way here," said Fenwick. He took a chair by the desk and glanced at the file folder, reading the title, Clearwater College. "And you've been hoping my application would get lost, and the whole thing would just disappear."

"Now, look, John—" Baker took his own seat behind the desk. Fenwick had always had a devilish knack for making him feel uncomfortable.

"It's all right," said Fenwick, waving away Baker's protests with a vigorous flap of his hand. "I know Clearwater isn't MIT or Cal Tech, but we've got a real hot physics department, and you're going to see some sparks flying out of there if you'll give us half a chance in the finance department. What's the good word, anyway? Do we get the research grant?"

Baker took a deep breath and settled his arms on the desk in front of him, leaning on them for support. He wished Fenwick wasn't so abrupt about things.

"John," Baker said slowly. "The head of your physics department doesn't even have a Ph. D. degree."

Fenwick brightened. "He's working on that, though! I told you that in answer to the question in the application. Bill, I wish you'd come down and see that boy. The things he can do with crystals would absolutely knock your hat off. He can stack them just like a kid stacking building blocks—crystals that nobody else has ever been able to manipulate so far. And the electrical characteristics of some of them—you wouldn't believe the transistors he's been able to build!"

"John," said Baker patiently. "The head of the physics department in any institution receiving a grant must have a Ph. D. degree. That is one absolutely minimum requirement."

"You mean we've got to wait until George finishes his work for his degree before we get the grant? That puts us in kind of a predicament because the work that we hoped to have George do under the grant would contribute towards his degree. Can't you put it through on the basis that he'll have his degree just as soon as the present series of experiments is completed?"

Baker wiped his forehead and looked down at his hands on the desk. "I said this is *one* minimum requirement. There are others, John."

"Oh, what else are we lacking?" Fenwick looked crestfallen for the first time.

"I may as well be blunt," said Baker. "There is no conceivable way in which Clearwater College can be issued a research grant for *anything*—and especially not for basic research in any field of physical science."

Fenwick just stared at him for a minute as if he couldn't believe what he had heard, although it was the thing he had expected to hear since the moment he sat down.

He seemed deflated when he finally spoke. "I don't think it was the intent of the Congressional Act that made these funds available," he said, "that only the big, plush outfits should get all the gravy. There are plenty of smaller schools just like Clearwater who have first rate talent in their science departments. It isn't fair to freeze us out completely—and I don't think it's completely legal, either."

"Clearwater is not being frozen out. Size has nothing to do with the question of whether an institution receives a grant from NBSD or not."

"When did you last give a grant to a college like Clearwater?"

"I am afraid we have never given a grant to a college—like Clearwater," said Baker carefully.

Fenwick's face began to grow more ruddy. "Then will you tell me just what is the matter with Clearwater, that we can't get any Government research contract when every other Tom, Dick, and Harry outfit in the country can?"

"I didn't state my case in exactly those terms, John, but I'll be glad to explain the basis on which we judge the qualifications of an institution to receive a grant from us."

Baker had never done this before for any unsuccessful applicant. In fact, it was the policy of the Bureau to keep the mysteries of the Index very carefully concealed from the public. But Baker wanted Fenwick to know what had hung him. It was the one more or less merciful thing he could do to show Fenwick what was wrong, and might be sufficient to shake him loose from his dismal association with Clearwater.

Baker opened the file folder and Fenwick saw now that it was considerably fuller than he had first supposed. Baker turned the pages, which were fastened to the cover by slide fasteners. Chart after chart, with jagged lines and multicolored areas, flipped by under Baker's fingers. Then Baker opened the accordian folds of a four-foot long chart and spread it on the desk top.

"This is the Index," he said, "a composite of all the individual charts which you saw ahead of it. This Index shows in graphical form the relationship between the basic requirements for obtaining a research grant and the actual qualifications of the applicant. This line marks the minimum requirement in each area."

Baker's finger pointed to a thin, black line that crossed the sheet. Fenwick observed that most of the colored areas and bars on the chart were well inside the area on Baker's side of the line. He guessed that the significance of the chart lay in this fact.

"I take it that Clearwater College is in pretty sad shape, chartwise," said Fenwick.

"Very," said Baker.

"Can you tell me how these charts are compiled?"

Baker turned back to the sheaf of individual charts. "Each item of data, which is considered significant in evaluating an applicant, is plotted individually against standards which have been derived from an examination of all possible sources of information."

"Such as?"

"For example, the student burden per faculty Ph. D. That is shown on this chart here."

"The what? Say that again," said Fenwick in bewilderment.

"The number of students enrolled, plotted against the number of doctorate degrees held by the faculty."

"Oh."

"As you see, Clearwater's index for this factor is dismally low."

"We're getting a new music director next month. She expects to get her doctorate next summer."

"I'm afraid that doesn't help us now. Besides, it would have to be in a field pertinent to your application to have much weight."

"George—"

"Doesn't help you at all for the present. You would require a minimum of two in the physics department alone. These two would have to be of absolutely top quality with a prolific publication record. That would bring this factor to a bare minimum."

"You take the number of Ph. D.'s and multiply them by the number of papers published and the years of experience and divide by the number of students enrolled. Is that the idea?"

"Roughly," said Baker. "We have certain constants which we also inject. In addition, we give weight to other factors such as patents applied for and granted. Periods of consultation by private industry, and so on. Each of these factors is plotted separately, then combined into the overall Index."

Baker turned the pages slowly, showing Fenwick a bleak record of black boundary lines cutting through nearly virginal territory on the charts. Clearwater's evaluation was reflected in a small spot of color near the bottom edge.

Fenwick stared at the record without expression for a long time. "What else do you chart?" he said finally.

"The next thing we evaluate is the performance of students graduated during the past twenty-five years."

"Clearwater is only ten years old," said Fenwick.

"True," said Baker, "and that is why, I believe, we have obtained such an anomalous showing in the chart of this factor."

Fenwick observed that the colored area had made a considerable invasion on his side of the boundary on this chart. "Why anomalous? It looks like we make a pretty good showing here."

"On the face of it, this is true," Baker admitted. "The ten-year record of the graduates of Clearwater is exceptional. But the past decade has been unusual in the scope of opportunities, you must admit."

"Your standard level must take this into account."

"It does. But somehow, I am sure there is a factor we haven't recognized here."

"There might be," said Fenwick. "There might be, at that."

"Another factor which contributes to the Index," said Baker, "is the cultural impact of the institution upon the community. We measure that in terms of the number and quality of cultural activities brought into the community by the university or college. We include concerts, lectures, terpsichorean activities, Broadway plays, and so on."

"Terpsichorean activities. I like that," said Fenwick.

"Primarily ballet," said Baker.

"Sure."

"Clearwater's record here is very low. It fact, there isn't any."

"This helps us get turned down for a research grant in physics?"

"It's a factor in the measurement of the overall status."

"Look," said Fenwick, "the citizens of Clearwater are so infernally busy with their own shindigs that they wouldn't know what to do if we brought a long-hair performance into town. If it isn't square-dancing in the Grange Hall, it's a pageant in the Masonic Temple. The married kids would probably like to see a Broadway play, all right, but they're so darned busy rehearsing their own in the basement of the Methodist Church that I doubt they could find time to come. Besides that, there's the community choir every Thursday, and the high school music department has a recital nearly every month. People would drop dead if they had any more to go to in Clearwater. I'd say our culture is doing pretty good."

"Folk activities are always admirable," said Baker, "but improvement of the cultural level in any community depends on the injection of outside influences, and this is one of the functions of the university. Clearwater College has not performed its obligation to the community in this respect."

Fenwick appeared to be growing increasingly ruddy. Baker thought he saw moisture appearing on Fenwick's forehead.

"I know this is difficult to face," said Baker sympathetically, "but I wanted you to understand, once and for all, just how Clearwater College appears to the completely objective eye."

Fenwick continued to stare at him without comment. Then he said flatly, "Let's see some more charts, Bill."

"Museum activities. This is an important function of a college level institution. Clearwater has no museum."

"We can't afford one, in the first place. In the second place, I think you've overlooked what we do have."

"There *is* a Clearwater museum?" Baker asked in surprise.

"Two or three hundred of them, I guess. Every kid in the county has his own collection of arrowheads, birds' eggs, rocks, and stuffed animals."

"I'm not joking, John," said Baker bleakly. "The museum aspect of the college is extremely important."

"What else?" said Fenwick.

"I won't go into everything we evaluate. But you should be aware of several other factors pertaining to the faculty, which are evaluated. We establish an index of heredity for each faculty member. This is primarily an index of ancestral achievement."

Fenwick's color deepened. Baker thought it seemed to verge on the purple. "Should I open the window for a moment?" Baker asked.

Fenwick shook his head, his throat working as if unable to speak. Then he finally managed to say, "Apart from the sheer idiocy of it, how did you obtain any information in this area?"

Baker ignored the comment, but answered the question. "You filled out forms. Each faculty member filled out forms."

"Yeah, that's right. I remember. Acres of forms. None of us minded if it was to help get the research grant. We supposed it was the usual Government razzmatazz to keep some GS-9 clerk busy."

"Our forms are hardly designed to keep people busy. They are designed to give us needed information about applicant institutions."

"And so you plot everybody's heredity."

"As well as possible. You understand, of course, that the data are necessarily limited."

"Sure. How do our grandpas stack up on the charts?"

"Not very well. Among Clearwater's total faculty of thirty-eight there were no national political figures through three generations back. There was one mayor, a couple of town councilmen, and a state senator or two. That is about all."

"Our people weren't very politically minded."

"This is a measure of social consciousness and contemporary evaluation."

Fenwick shrugged. "As I said, we aren't so good at politics."

"Achievements in welfare activities are similarly lacking. No notable intentions or discoveries, with the exception of one patent on a new kind of beehive, appear in the record."

"And this keeps us from getting a research grant in physics? What *did* our progenitors do, anyway? Get hung for being horse thieves?"

"No criminal activities were reported by your people, but there is a record of singular restlessness and dissatisfaction with established conditions."

"What did they do?"

"They were constantly on the move, for the most part. In the eighteenth and nineteenth centuries they were primarily pioneers, frontiersmen, settlers of new country. But when the country was established they usually packed up and went somewhere else. Rovers, trappers, unsettled people."

"This is not good?" Fenwick glanced at the chart that was open now. It was almost uncolored.

"I regret to say that such people are not classed as the stable element of communities," said Baker. "We cannot evaluate the index of hereditary accomplishment for the Clearwater faculty very high."

"It appears that our grandpas were among those generally given credit for getting things set up," said Fenwick.

"Such citizens are indeed necessary," said Baker. "But our index evaluates stability in community life and accomplishments with long-range effects in science and culture."

"We haven't got much of a chance then, grandpa being foot-loose as he was."

"Other factors could completely override this negative evaluation. You see, this is the beauty of the Index; it doesn't depend on any one factor or small group of factors. We evaluate the whole range of factors that have anything to do with the situation. Weaknesses in one spot may be counterbalanced by strength in others."

"It looks like Clearwater is staffed by a bunch of bums without any strong spots."

"I wouldn't say it in such terms, but the reason I am pointing these things out to you, John, is to try to persuade you to disassociate yourself from such a weak organization and go elsewhere. You have fine talents of your own, but you have always had a pattern of associating with groups like this one at Clearwater. Don't you see now that the only thing for you to do is go somewhere where there are people capable of doing things?"

"I *like* Clearwater. I like the people at the College. Where else are we in the bums category?"

Baker suddenly didn't want to go on. The whole thing had become distasteful to him. "There are a good many others. I don't think we need to go into them. There is the staff reading index, the social activity index, wardrobe evaluation, hobbies, children—actual and planned."

"I want to hear about them," said Fenwick. "That wardrobe evaluation—that sounds like a real fascinating study."

"Actually, it's comparatively minor," said Baker. "Our psychologists have worked out some extremely interesting correlations, however. Each item of a man's wardrobe is assigned a numerical rating. Tuxedo, one or more. Business suits, color and number. Hunting jackets. Slacks. Sport coats. Work shoes. Dress shoes. Very interesting what our people can do with, such information."

"Clearwater doesn't rate here?"

Baker indicated the chart. "I'm afraid not. Now, this staff reading index is somewhat similar. You recall the application forms asked for the number of pages of various types of material read during the past six months—scientific journals, newspapers, magazines, fiction."

"I suppose Clearwater is a pretty illiterate bunch," said Fenwick.

Baker pointed soundlessly to the graph.

"Hobbies and social activities are not bad," Baker said, after a time. "Almost up to within ten points of the standard. A few less bingo parties and Brownie meetings and that many more book reviews or serious soirees would balance the social activity chart. If the model railroad club were canceled and a biological activity group substituted, the hobby classification would look much better. Then, in the number of children, actual and planned, Clearwater is definitely out of line, too. You see, the standard takes the form of the well-known bell-shaped curve. Clearwater is way down on the high side."

"Too much biological activity already," Fenwick murmured.

Baker looked up. "What was that? I didn't hear what you said."

Fenwick leaned back and extended his arms on the desk. "I said your whole damned Index is nothing but a bunch of pseudo-intellectual garbage."

Baker felt the color rising in his face, but he forced himself to remain calm. After a moment of silence he said. "Your emotional feelings are understandable, but you must remember that the Index permits us to administer accurately the National Science Development Act. Without the scientific assurance of the Index there would be no way of determining where these precious funds could best be utilized."

"You'd be better off putting the money on the ponies," said Fenwick. "Sometimes they win. As it stands, you've set it up for a sure loss. You haven't got a chance in the world."

"You think Clearwater College could make better use of some of our funds than, say, MIT?"

"I wouldn't be surprised. Don't get me wrong. I'm not saying the boys at MIT or Cal Tech or a lot of other places couldn't come up with a real development in the way of a fermodacular filter for reducing internucleated cross currents. But the real breakthroughs—you've closed your doors and locked them out."

"Who have we locked out? We've screened and fine combed the resources of the entire country. We know exactly where the top research is being conducted in every laboratory in the nation."

Fenwick shook his head slowly and smiled. "You've forgotten the boys working in their basements and in their back yard garages. You've forgotten the guys that persuade the wife to put up with a busted-down automatic washer for another month so they can buy another hundred bucks worth of electronic parts. You've remembered the guys who have Ph. D.'s for writing 890-page dissertations on the Change of Color in the Nubian Daisy after Twilight, but you've forgotten guys like George Durrant, who can make the atoms of a crystal turn handsprings for him."

Baker leaned back in his chair and smiled. He almost wished he hadn't wasted the effort of trying to show Fenwick. But, then, he had tried. And he would always have regretted it if he hadn't.

"You're referring now to the crackpot fringe?" he said.

"I suppose so," said Fenwick. "I've heard it called that before."

"One of the things, above all else, which the Index was designed to accomplish," said Baker, "was the screening out of all elements that might be ever so remotely associated with the crackpot fringe. And believe me, you'll never know how strong it is in this country! Every two-bit tinkerer wants a handout to develop his world-shaking gadget that will suppress the fizz after the cap is removed from a pop bottle, or adapt any apartment-size bathtub for raising tropical fish."

"You ever heard of the flotation process?" said Fenwick abruptly.

Baker frowned at the sudden shift of thought. "Of course—"

"What would the world be like without the flotation process?"

"The metals industry would be vastly different, of course. Copper would be much scarcer and higher priced. Gold—"

"A ton of ore and maybe a pound of recovered metal, right?" said Fenwick. "Move a mountain of waste to get anything of value. Crush millions of tons of rock and float out the pinpoint particles of metal on bubbles of froth."

"That's a rough description of what happens."

"You've heard of high-grading."

"Of course. A somewhat colloquial term used in mining."

"The high-grader takes a pick and digs for anything big enough to see and pick up with his hands. He doesn't worry about the small stuff that takes sweat and machinery to recover."

"I suppose so. I fail to see the significance—"

"You're high-grading, Bill," said Fenwick. He leaned across the desk and spoke with bitter intensity. "You're high-grading and you should be using a flotation process."

Fenwick slowly drew back in his chair. Baker felt overwhelmed by the sudden intensity he had never before seen displayed in John Fenwick. Any reaction on his part seemed suddenly inadequate. "I fail to see any connection—," he said finally.

Fenwick looked at him steadily. "Human creativeness can be mined only by flotation methods. It's in low-grade ore. Process a million stupid notions and find a pin point of genius. Turn over enormous wastes of human thought and recover a golden principle. But turn your back on these mountains of low-grade material and you shut out the wealth of creative thought that is buried in them. More than that, by high-grading only where rich veins have appeared in the past, you're mining lodes that have played out."

"An ingenious analogy," said Baker, recovering with a smile now. "But it's hardly an accurate or applicable one. The human mind is not a piece of precious metal found in a mountain of ore. Rather, it's an intricate device capable of producing computations of unbelievable complexity. And we know how such devices that are superior in function are produced, and we know what their characteristics are. We also know that such a device does not 'play out'. If it is superior in function, it can remain so for a long time."

"High-grading," said Fenwick. "And the vein is played out. You'll never find the thing you're looking for until you develop means of processing low-grade material."

Baker watched Fenwick across the desk. He was weary of the whole thing. He certainly had no need to prove himself to this man. He had simply tried to do Fenwick a favor, and Fenwick had thrown it right back in his face. Yet there was a temptation to go on, to prove to Fenwick the difference between their two worlds. Fenwick belonged to a world compounded of inevitable failure. The temptation to show him, to try again to lift him out of it was born of a kind of pity for Fenwick.

Baker's own life had arrowed decisively, without waver, to a goal that was as correct as the tolerances of human error could make it. He often permitted himself the pride of considering his mind somewhat as a computer that had been programmed through a magnificent gene inheritance to drive irresistibly toward the precise goals he had reached. But Fenwick—Fenwick was still fumbling around in a morass of uncertainty. After years of erratic starts and stops he was now confusedly trying to make something out of that miserable little institution called Clearwater College.

It wasn't particularly friendship that urged Baker to show Fenwick. Their friendship was of a breed that Baker had never quite been able to define to his own satisfaction. It seemed to him there was a sort of deadly fascination in associating with a man who walked so blindly, who was so profoundly incapable of understanding his own blindness and peril.

"I'm going to show you," Baker said abruptly, "exactly what it would mean if we were to do as you suggest. I'll show you what it would be like to give attention to every halfwit and crackpot that comes begging for a handout." He switched the intercom and spoke into it. "Doris, please bring in the Ellerbee file. Yes—the crackpot section."

He switched off. "Doris has her own quaint but quite accurate way of cataloguing our various applications," he explained.

In a moment the secretary entered and placed the file on the desk. "There's a new letter in there," she said. "Dr. Pehrson initialed it. He said you didn't want to be bothered any more with this case."

"That's right."

Baker opened the file and shoved it toward Fenwick. "This boy has a gadget he wants us to look at. Doesn't really need any money, he says. That's the kind we really have to be on guard against. If we looked at his wonder gadget, we'd be pestered for a million-dollar handout for years to come."

"What's he got?" Fenwick asked.

"Some kind of communication device, he says. He claims it's nothing but a grown crystal which you hold in your hand and talk to anybody anywhere on Earth."

"Sounds like it wouldn't take much to find out whether he's got anything or not. Just let him put on a five-minute demonstration."

"But multiply that five minutes by a thousand, by ten thousand. And once you let them get their teeth into you, it doesn't stop with five minutes. It goes on into reams of letters and years of time. No, you have to stop this kind of thing before it ever starts. But take a look at some of this material in the file and you'll see what I mean."

Fenwick picked up the top letter as Baker pushed the file toward him. "He starts this one by saying, 'Dear Urban.' Is that what he calls you? What does he mean?"

"Who knows? He's a crackpot, I told you. Who cares what he means, anyway. We've got far more important things to worry about."

Fenwick scanned the letter a moment, then looked up, a faint smile on his face. "I know what he means. Urban—Pope Urban—was the one responsible for the persecutions of Galileo."

Baker shrugged embarrassedly. "I told you he was a crackpot. Delusions of grandeur and of persecution are typical."

"This boy may not be as crazy as he sounds. You're giving him a pretty good imitation of a Galileo treatment—won't even look at his device. He says here that 'Since you have previously refused to examine my device and have questioned my reliability as an observer, I have obtained the services of three unbiased witnesses, whose affidavits, signed and notarized, are attached. These men are the Fire Chief, the Chief of Police, and the Community Church Pastor of Redrock, all of whom testify that they did see my device in full operation this past week. I trust that this evidence will persuade you that an investigation should be made of my device. I fail to see how the bull-headedness and cocksureness of your office can withstand any more of the evidence I have to offer in support of my claims.'"

"A typical crackpot letter," said Baker. "He tries to be reasonable, but his colors are soon shown when he breaks down into vituperative language like a frustrated child."

Fenwick thumbed through the large pile of correspondence. "I'd say anybody would likely blow his stack a good deal harder than this if he'd been trying to get your attention this long. Why didn't he ever send you one of his gadgets in the mail?"

"Oh, he did," said Baker. "That was one of the first things he did."

"What did you do?"

"Sent it back. We always return these things by registered return mail."

"Without even trying it out?"

"Of course."

"Bill, that isn't even reasonable. These earlier letters of his describe the growing of these crystals. He tells exactly how he does it. He knows what he's talking about. I'd like to see him and see his crystal."

"That's what I was hoping you'd say! All we have to do is get Doris to give him a call and he'll be here first thing in the morning. You can be our official investigator. You can see what it's like dealing with a crackpot!"

James Ellerbee was a slim man, impetuous and energetic. Fenwick liked him on sight. He was not a technical man; he was a farmer. But he was an educated farmer. He had a degree from the State Agricultural College. He dabbled in amateur radio and electronics as a hobby.

"I'm certainly glad someone is finally willing to give me a break and take a look at my device," he said as he shook Fenwick's hand. "I've had nothing but a runaround from this office for the past eight months. Yet, according to all the publicity, this is where the nation's scientific progress is evaluated."

Fenwick felt like a hypocrite. "We get pretty overloaded," he said lamely.

They were in Baker's office. Baker watched smugly from behind his desk. Ellerbee said, "Well, we might as well get started. All you have to do, Mr. Fenwick, is hold one of these crystal cubes in your hand. I'll go in the other office and close the door. It may help at first if you close your eyes, but this is not really necessary."

"Wait," said Fenwick. Somehow he wanted to get away from Baker while this was going on. "I'd like to take it outside, somewhere in the open. Would that be all right?"

"Sure. Makes no difference where you try it," said Ellerbee. "One place is as good as another."

Baker waved a hand as they went out. "Good luck," he said. He smiled confidently at Fenwick.

As far as Fenwick could see, the crystal was not even potted or cased in any way. The raw crystal lay in his hand. The striations of the multitude of layers in which it was laid down were plainly visible.

Ellerbee dropped Fenwick off by the Jefferson memorial, then drove on about a mile. Still in sight, he stopped the car and got out. Fenwick saw him wave a hand. Nothing happened.

Fenwick glanced down at the crystal in his hand. About the size of a child's toy block. He could almost understand Baker's position. It *was* pretty silly to suppose this thing could have the powers Ellerbee said it had. No electric energy applied. It merely amplified the normal telepathic impulses existing in every human mind, Ellerbee said. Fenwick sighed. You just couldn't tell ahead of time that a thing wasn't going to pan out. He knew his philosophy was right. These had to be investigated—every lousy, crackpot one of them. You could never tell what you were missing out on unless you did check.

He squeezed harder on the crystal, as Ellerbee had told him to do.

It was just a little fuzzy at first, fading and coming back. Then it was there, shimmering a little, but steady. The image of Ellerbee standing in front of him, grinning.

Fenwick glanced down the road. Ellerbee was still there, a mile away. But he was also right there in front of him, about four feet away.

"It shakes you up a little bit at first," said Ellerbee. "But you get used to it after a while. Anyway, this is it. Are you convinced my device works?"

Fenwick shook his head to try to clear it rather than to give a negative answer. "I'm convinced *something* is working," he said. "I'm just not quite sure what it is."

"I'll drive across town," Ellerbee offered. "You can see that distance makes no difference at all. Later, I'll prove it works clear across the country if you want me to."

They arranged that proof of Ellerbee's presence on the other side of the city could be obtained by Fenwick's calling him at a drug store pay phone. Then they would communicate by means of the cubes.

It was no different than before.

The telephone call satisfied Fenwick that Ellerbee was at least ten miles away. Then, within a second, he also appeared to be standing directly in front of Fenwick.

"What do you want?" said Fenwick finally. "What do you want the Bureau to do about your device? How much money do you want for development?"

"Money? I don't need any money!" Ellerbee exploded. "All I want is for the Government to make some use of the thing. I've had a patent on it for six months. The Patent Office had sense enough to give me a patent, but nobody else would look at it. I just want somebody to make some use of it!"

"I'm sure a great many practical applications can be found," Fenwick said lamely. "We'll have to make a report, first, however. There will be a need for a great many more experiments—"

But most important of all, Baker would have to be shown. Baker would have to *know* from his own experience that this thing worked.

Fenwick suddenly wanted to get away from Ellerbee as much as he had from Baker a little earlier. There was just so much a man's aging synapses could stand, he told himself. He had to do a bit of thinking by himself. When Ellerbee drove up again, Fenwick told him what he wanted.

Ellerbee looked disappointed but resigned. "I hope this isn't another runaround, Mr. Fenwick. You'll pardon me for being blunt, but I've had some pretty raw treatment from your office since I started writing about my communicator."

"I promise you this isn't a runaround," said Fenwick, "but it's absolutely necessary to get Dr. Baker to view your demonstration. We will want to see your laboratories and your methods of production. I promise you it won't be more than two or three days, depending on Dr. Baker's busy schedule."

"O.K. I'll wait until the end of the week," said Ellerbee. "If I don't hear something by then, I'll go ahead with my plans to market the crystals as a novelty gadget."

"I'll be in touch with you. I promise," said Fenwick. He stood by the curb and watched Ellerbee drive away.

Fenwick moved slowly back to his own car and sat behind the wheel without starting the motor. It seemed a long time since nine-thirty yesterday morning, when he had come in to Baker's office to check on the grant he had known Baker wasn't going to give him. Now, merely by kicking Baker's refuse pile with his toe, so to speak, he had turned up a diamond that Baker was ready to discard.

Fenwick felt a sudden surge of revulsion. How was it possible for such a blind, ignorant fool as Baker to be placed in the position he was in? How could the administrative officers of the United States Government be responsible for such misjudgment? Such maladministration, if performed consciously, would be sheer treason. Yet, unconsciously and ignorantly, Baker's authority was perpetuated, giving him a stranglehold on the creative powers of the nation.

Fenwick tried to recall how he and Baker had become friends—so long ago, in their own college days. It wasn't that there was any closeness or common interest between them, yet they seemed to have drawn together as two opposites might. They were both science majors at the time, but their philosophies were so different that their studies were hardly a common ground.

Fenwick figuratively threw away the textbook the first time the professor's back was turned. Baker, Fenwick thought, never took his eyes from its pages. Fenwick distrusted everything that he could not prove himself. Baker believed nothing that was not solidly fixed in black and white and bound between sturdy cloth covers, and prefaced by the name of a man who boasted at least two graduate degrees.

Fenwick remembered even now his first reaction to Baker. He had never seen his kind before and could not believe that such existed. He supposed Baker felt similarly about him, and, out of the strange contradiction of their worlds, they formed a hesitant friendship. For himself, Fenwick supposed that it was based on a kind of fascination in associating with one who walked so blindly, who was so profoundly incapable of understanding his own blindness and peril.

But never before had he realized the absolute danger that rested in the hands of Baker. And there must be others like him in high Government scientific circles, Fenwick thought. He had learned long ago that Baker's kind was somewhere in the background in every laboratory and scientific office.

But few of them achieved the strangling power that Baker now possessed.

The Index! Fenwick thought of it and gagged. Wardrobe evaluation! Staff reading index! The reproductive ratio—social activity index—the index of hereditary accomplishment—multiply your ancestors by the number of technical papers your five-year old children have produced and divide by the number of book reviews you attend weekly—

Fenwick slumped in the seat. We hold these truths to be self-evident—that the ratio of sports coats to tuxedos in a faculty member's closet shall determine whether Clearwater gets to do research in solid state physics, whether George Durrant gives his genius to the nation or whether it gets buried in Dr. William Baker's refuse pile.

But not only George Durrant. Jim Ellerbee, too. And how many others?

Something had to be done.

Fenwick hadn't realized it before, but this was the thought that had been churning in his cortex for the last hour. Something had to be done about Bill Baker.

But, short of murder, what?

Getting rid of Baker physically was not the answer, of course. If he were gone, a hundred others like him would fight for his place.

Baker had to be shown. He had to be shown that high-grading was costing him the very thing he was trying to find. It must be proven to him that flotation methods work as well in mining human resources as in mining metal. That the extra trouble paid off.

This was known—a long time ago—Fenwick thought. Somewhere along the way things got changed. He glanced toward the Jefferson Memorial. Tom Jefferson knew how it should be, Tom Jefferson, statesman, farmer, writer, and amateur mechanic and inventor. It was not only every gentleman's privilege, it was also his duty to be a tinkerer and amateur scientist, no matter what else he might be.

Fenwick glanced in the distance toward the Lincoln Memorial. Abe had done his share of tinkering. His weird boot-strap system for hoisting river boats off shoals and bars hadn't amounted to much, but Abe knew the principle that every man has the right to be his own scientist.

And then there was Ben Franklin, the noblest amateur of them all! He had roamed these parts, too.

Somewhere it had been lost. The Bill Bakers would have laughed out of existence the great tinkerers like Franklin and Lincoln and Jefferson. And the Pasteurs and the Mendels—and the George Durrants and the Jim Ellerbees, too.

Fenwick started the car. Something had to be done about Bill Baker.

Baker leaned back in his chair and laughed heartily. "So it worked, did it? He showed you something that made you think he had a real working device."

"There was no 'think' about it," said Fenwick. "I saw it with my own eyes. That boy's got something terrific!"

Baker sobered and thumbed through the Ellerbee file again. "Any freshman math major could poke holes all through this mathematical explanation he offers. Right? Secondly, a device such as he claims to have produced violates all the basic laws of science. Why, it's even against the Second Law of Thermodynamics!"

"I don't care what it's against," said Fenwick. "It works. I want you to come with me to Ellerbee's and see for yourself. His device will revolutionize communications."

Baker shook his head sadly. "It's always tougher when they show you something that seems to work. Then you've got to waste a lot of time looking for the gimmick if you're going to follow it through. I just haven't got the time—"

"You've got to, Bill!"

"I'll tell you what I'll do. You go out there and look over his setup. If you can't find his gimmick in half a day, I'll come out and show it to you. But I warn you, some of these things are very tricky—like the old perpetual motion machines. You've got to have your wits about you. Is that fair enough?"

"Fair enough," Fenwick agreed.

Baker smiled broadly. "I'll do even more. If this Ellerbee device should prove to be on the level, I'll give you the research grant you want for Clearwater."

"I'm not so sure I want it on those terms," said Fenwick.

"Well, it's a purely academic matter. You won't have to worry about it. But, on the other hand, I'll expect you to agree that when Ellerbee is exposed you'll not persist in your request to this office."

"Well, now—"

"That's a fair offer. I'm giving you a chance to prove I'm wrong in setting up the Index to screen out people like Ellerbee—"

"—And institutions like Clearwater."

"And institutions like Clearwater," Baker agreed.

"All right," said Fenwick. "I'll gamble with you—for one more stake: If Ellerbee's device is on the level, you'll make a grant to Clearwater *and* other institutions of like qualifications, and you'll scrap that insane Index—"

Baker tapped the desk placatingly. "The grant to Clearwater, yes. As for the Index, if it should fail in its applicability to this clear-cut Ellerbee case I would be the first to want to know why. But I assure you there is no flaw in the Index. It has been tried too many thousands of times."

Ellerbee's place was in Virginia, in a dairying area in the hills. The last ten miles of the road were not the kind to attract visitors. The road was steep and narrow in places that turned sharply around the hillsides. No guardrails blocked the descent into the steep gullies. It was definitely a region for people who liked solitude. The farms that lay in the valleys of the hills were neat and well-cared for, however. The people Fenwick passed on the road didn't look like the recluse type.

Ellerbee's farm was one of the best looking in the vicinity. It had the look of being cared for by a man who could do everything. The huge barn and the corrals were as neat as a garden, and the large white frame farmhouse stood out like a monument against the green pasture.

A woman and two children were in the garden beside the house as Fenwick drove up. "May I help you? I'm Mrs. Ellerbee," the woman said.

Fenwick explained who he was and his purpose in coming. "Jim's been expecting you," the woman said. "His laboratory is the long white building back of the house. He's out there now."

Jim Ellerbee met him at the door. "You didn't bring Dr. Baker," he said almost accusingly.

"Later," said Fenwick. "I came, as I promised. Dr. Baker wants my report on your facilities and production methods. Then he will come up to make his own inspection."

There was doubt in Ellerbee's eyes, as if he was used to such stories. "Maybe it would be best if I marketed the crystals in any form I can," he said.

He led Fenwick through a number of rooms of expensive, precision electronic equipment. Then they passed through a set of double doors, which Fenwick observed acted as a thermal lock between the crystal growing room and the rest of the building. It reminded him of George Durrant's laboratory at Clearwater.

"This is where the crystals are grown," said Ellerbee. "I suppose you're familiar with such processes. Here we must use a very precisely controlled sequence of co-crystallization to get layers of desired thickness—"

Fenwick wasn't listening. He had suddenly observed the second man in the room, a rather small, swarthy man, who moved with quiet precision among a row of tanks on the far side of the room. There was a startling quality about the man that Fenwick was unable to define, a strangeness.

Ellerbee caught the direction of his glance. "Oh," he said. "You must meet my neighbor, Sam Atkins. Sam is in this as deep or even deeper than I am. I think perhaps he's more responsible for the communicator crystals."

The man turned as his name was mentioned, and came toward them. "You were the one who developed the crystals," he said in a soft, persuasive voice, to Jim Ellerbee.

"This is my setup," Ellerbee explained with a wave of his hand to indicate the laboratory surroundings. "But Sam has been working with me for about a year on this thing. When Sam moved in, we found we were both radio hams and electronic bugs. I'd been fooling around with crystal growing, trying to design some new type transistors. Then Sam suggested some experiments in co-crystallization—using different chemicals that will crystallize in successive layers in one crystal.

"We stumbled on one combination that made a terrific amplifier. Then we found it would actually radiate to a distant point all by itself. Finally, we discovered that its radiation was completely nonelectromagnetic. There is no way we have yet found of detecting the radiation from the crystal—except by means of another piece of the same crystal.

"I know it's against all the rules in the books. It just doesn't make sense. But there it is. It works."

Sam Atkins had turned away for a moment to attend to one of the tanks, but Fenwick found himself intensely aware of the man's presence. There was nothing he could put his finger on. He just knew, with such intense certainty, that Sam Atkins was *there*.

"What does Mr. Atkins do?" Fenwick asked. "Does he have a dairy farm, too?"

Ellerbee nodded. "His place is right next to mine. Since we started this project Sam has practically lived here, however. He's a bachelor, and so he takes most of his meals with us."

"Seems strange—" Fenwick mused, "two men like you, way out here in the country, doing work on a level with that of the best crystal labs in the country. I should think you'd both rather be in academic or industrial work."

Ellerbee smiled and looked up through the windows to the meadows beyond. "We're *free* out here," he said.

Fenwick thought of Baker. "You are that," he said.

"You said you wanted to investigate the whole production process. We'll start here, if you like, and I'll show you every step in our process. This tank contains an ordinary alum solution. We start building on a seed crystal of alum and continue until we reach a precise thickness. Here is a solution of chrome alum. You'll note the insulated tanks. Room temperature is maintained within half a degree. The solutions are held to within one-tenth of a degree. Crystal dimensions must be held to tolerances of little more than the thickness of a molecule—"

The gimmick to fool him and cheat him. Where was it? Fenwick asked himself. Baker was sure it was here. If so, where could it be? There was no trickery in the crystal laboratory—unless it was the trickery of precision refinement of methods. Only men of great mechanical skill could accomplish what Ellerbee and his friend were doing. Genius behind the milking machine! Fenwick could almost sympathize with Baker in his hiding behind the ridiculous Index. Without some such protection a man could encounter shocks.

The crackpot fringe.

Where else would credence have been given to the phenomenon of a crystal that seemed to radiate in a nonelectromagnetic way?

But, of course, it couldn't actually be doing that. All the books, all the authorities, and the known scientific principles said it couldn't happen. Therefore, it wouldn't have happened—outside the crackpot fringe.

If Ellerbee and Atkins weren't trying to foist a deliberate deception, where were they mistaken? It was possible for such men as these to make an honest mistake. That would more than likely turn out to be the case here. But how could there be a mistake in the production of a phenomenon such as Fenwick had witnessed? How could that be produced through some error when it couldn't even be done by known electronic methods—not just as Fenwick had seen it.

Throughout the morning Ellerbee led him down the rows of tanks, explaining at each step what was happening. Sometimes Sam Atkins offered a word of explanation also, but always he stayed in the background. The two farmers showed Fenwick how they measured crystal size down to the thickness of a molecule while the crystals were growing.

A sudden suspicion crossed Fenwick's mind. "If those dimensions are so critical, how did you determine them in the first place?"

"Initially, it was a lucky accident," said Sam Atkins.

"Very lucky," said Fenwick, "if you were able to accidentally obtain a crystal of fifteen layers or so and have each layer even approximately correct."

Sam smiled blandly. "Our first crystals were not so complex, you understand. Only three layers. We thought we were building transistors, then. Later, our mathematics showed us the advantage of additional layers and gave us the dimensions."

The mathematics that Baker said a kid could poke holes in. Fenwick didn't know. He hadn't checked the math.

Where was the gimmick?

In the afternoon they took him out for field tests again. A rise behind the barn was about a mile from a similar rise on Sam Atkins' place. They communicated across that distance in all the ways, including various kinds of codes, that Fenwick could think of to find some evidence of hoax. Afterwards, they returned to the laboratory and sawed in two the crystals they had just used. Then they showed him the tests they had devised to determine the nature of the radiation between the crystals.

He did not find the gimmick.

By the end of the day Ellerbee seemed beat, as if he'd been under a heavy strain all day long. And then Fenwick realized that was actually the case. Ellerbee wanted desperately to have someone believe in him, believe in his communication device. Not only had he used all the reasoning power at his command, he had been straining physically to induce Fenwick to believe.

Through it all, however, Sam Atkins seemed to remain bland and utterly at ease, as if it made absolutely no difference to him, whatever.

"I guess we've just about shot our wad," said Ellerbee. "That's about all we've got to show you. If we haven't convinced you by now that our communicator works, I don't know how we can accomplish it."

Had they convinced him? Fenwick asked himself. Did he believe what he had seen or didn't he? He had been smug in front of Baker after the first demonstration, but now he wondered how much he had been covered by the same brush that had tarred Baker.

It wasn't easy for him to admit the possibility of nonelectromagnetic radiation from these strange crystals, radiation which could carry sight and sound from one point to another without any transducers but the crystals themselves.

"You have to step out of the world you've grown accustomed to," said Sam Atkins very quietly. "This is what we have had to do. It's not hard now to comprehend that telepathic forces of the mind can be directed by this means. This is a new pattern. Think of it as such. Don't try to cram it into the old pattern. Then it's easy."

A new pattern. That was the trouble, Fenwick thought. There couldn't really be any new patterns, could there? There was only one basic pattern, in which all the phenomena of the universe fit. He readily admitted that very little was known about that pattern, and many things believed true were false. But the Second Law of Thermodynamics. *That* had to be true—invariably true—didn't it?

If there was a hoax, Baker would have to find it.

"I'll be back with Dr. Baker in a couple of days," Fenwick said. "After that, the one final evidence we'll need will be to construct these crystals in our own laboratories, entirely on our own, based on your instructions."

Ellerbee nodded agreement. "That would suit us just fine."

"Hypnotism," said Baker. "It sounds like some form of hypnotism, and I don't like that kind of thing. It could merit criminal prosecution."

"There's no possible way I could have been hypnotized," said Fenwick.

"These crystals—obviously it has something to do with them. But I wonder what their game is, anyway? It's hard to see where they can think they're headed."

"I don't know," said Fenwick. "But you promised to show me the gimmick if I couldn't find it in half a day. I spent a whole day out there without finding anything."

Baker slapped the desk in exasperation. "You're not really going to make me go out there and look at this fool thing, are you? I know I made a crazy promise, but I was sure you could find where they were hoaxing you if you took one look at their setup. It's probably so obvious you just stumbled right over it without even seeing it was there."

"Possibly. But you're going to have to show me."

"John, look—"

"Or, I *might* be willing to take that Clearwater research grant without any more questions on either side."

Baker thought of the repercussions that would occur in his own office, let alone outside it, if he ever approved such a grant. "All right," he sighed. "You've got me over a barrel. I'll drive my car. I may have to stop at some offices on the other side of town."

"I might be going on, rather than coming back to town," said Fenwick. "I ought to have my car, too. Suppose I meet you out there?"

"Good enough. Say one o'clock. I'm sure that will give us more time than we need."

Baker was prompt. He arrived with an air of let's-get-this-over-as-quick-as-possible. He nodded perfunctorily as Ellerbee introduced his wife. He scarcely looked at Sam Atkins.

"I hope you've got your demonstration all set up," he said. He glanced at the darkening sky. "It looks like we might get some heavy rain this afternoon."

"We're all ready," said Ellerbee. "Sam will drive over to that little hill on his farm, and we'll go out behind the barn."

On the knoll, Baker accepted the crystal cube without looking at it. Clenching it in his fist, he put his hand in his pocket. Fenwick guessed he was trying to avoid any direct view and thus avoid the possibility of hypnotic effects. This seemed pretty farfetched to Fenwick.

Through field glasses Sam Atkins was seen to get out of his car and walk to the top of the knoll. He stood a moment, then waved to signal his readiness.

"Press the crystal in your hand," Ellerbee said to Baker. "Direct your attention toward Sam Atkins."

Each of them had a cube of the same crystal. It was like a party line. Fenwick pressed his only slightly. He had learned it didn't take much. He saw Baker hesitate, then purse his lips as if in utter disgust, and follow instructions.

In a moment the image of Sam Atkins appeared before them. Regardless of their position, the image gave the illusion of standing about four feet in front of them.

"Good afternoon, Dr. Baker," Sam Atkins said.

Fenwick thought Baker was going to collapse.

The director just stood for a moment, like a creature that had been pole-axed. Then his color began to deepen and he turned with robot stiffness. "Did you men hear anything? Fenwick ... did you hear ... did you see?"

"Sure," said Fenwick, grinning broadly. "Sam Atkins said good afternoon to you. It would be polite if you answered him back."

The image of Sam Atkins was still before them. He, too, was grinning broadly. The grins infuriated Baker.

"Mr. Atkins," said Baker.

"Yes, Dr. Baker," said Sam Atkins.

"If you hear me, wave your hands. I will observe you through the field glasses."

"The field glasses won't be necessary."

Both the image before them, and the distant figure on the knoll were seen to wave arms and gyrate simultaneously. For good measure, Sam Atkins turned a cartwheel.

Baker seemed to have partly recovered. "There seems to be a very remarkable effect present here," he said cautiously.

"Dr. Baker," Jim Ellerbee spoke earnestly, "I know you're skeptical. You don't think the crystals do what I say they do. Even though you see it with your own eyes you doubt that it is happening. I will do anything possible to test this device to your satisfaction. Name the test that will dispel your doubts and we will perform it!"

"It's not entirely a question of demonstration, Mr. Ellerbee," said Baker. "There are the theoretical considerations as well. The mathematics you have submitted in support of your claim are not, to put it mildly, sound."

"I know. Sam keeps telling me that. He says we need an entirely new math to handle it. Maybe we'll get around to that. But the important thing is that we've got a working device."

"Your mathematical basis *must* be sound!" Baker's fervor returned. Fenwick felt a sudden surge of pity for the director.

The demonstration was repeated a dozen times more. Fenwick went over on Sam Atkins' hill. He and Baker conversed privately.

"Do you see it yet?" Fenwick asked.

"No, I'm afraid I don't!" Baker was snappish. "This is one of the most devilish things I've ever come across!"

"You don't think it's working the way Jim and Sam say it is?"

"Of course not. The thing is utterly impossible! I am convinced a hypnotic condition is involved, but I must admit I don't see how."

"You may figure it out when you go through Ellerbee's lab."

Baker was obviously shaken. He spoke in subdued tones as Ellerbee started the tour of the crystal lab again. Baker's eyes took in everything. As the tour progressed he seemed to devour each new item with frenzied intensity. He inspected the crystals through a microscope. He checked the measurements of the thickness of the growing crystal layers.

The rain began while they were in the crystal lab. It beat furiously on the roof of the laboratory building, but Baker seemed scarcely aware that it was taking place. His eyes sought out every minute feature of the building. Fenwick was sure he was finding nothing to confirm his belief that the communicator crystals were a hoax.

Fenwick hadn't realized it before, but he recognized now that it would be a terrific blow to Baker if he couldn't prove the existence of a hoax.

Proof that the communicator crystals were all they were supposed to be would be a direct frontal attack on the sacred Index. It would blast a hole in Baker's conviction that nothing of value could come from the crackpot fringe. And, not least of all, it would require Baker to issue a research grant to Clearwater College.

What else it might do to Baker, Fenwick could only guess, but he felt certain Bill Baker would never be the same man again.

As it grew darker, Baker looked up from the microscope through which he had been peering. He glanced at the windows and the drenched countryside beyond. "It's been raining," he said.

Mary Ellerbee had already anticipated that the visitors would be staying the night. She had the spare room ready for Baker and Fenwick before dinner. While they ate in the big farmhouse kitchen, Ellerbee explained. "It would be crazy to try to get down to the highway tonight. The county's been promising us a new road for five years, but you see what we've got. Even the oldest citizen wouldn't tackle it in weather like this, unless it was an emergency. You put up for the night with us. You'll get home just as fast by leaving in the morning, after the storm clears. And it will be a lot more pleasant than spending the night stuck in the mud somewhere—or worse."

Baker seemed to accept the invitation as he ate without comment. To Fenwick he appeared stunned by the events of the day, by his inability to find a hoax in connection with the communicator crystals.

It was only when Baker and Fenwick were alone in the upstairs bedroom that Baker seemed to stir out of his state of shock.

"This is ridiculous, Fenwick!" he said. "I don't know what I'm doing here. I can't possibly stay in this place tonight. I've got people to see this evening, and appointments early in the morning."

"It's coming down like cats and dogs again," said Fenwick. "You saw the road coming in. It's a hog wallow by now. Your chance of getting through would be almost zero."

"It's a chance I have to take," Baker insisted. He started for the door. "*You* don't have to take it, of course."

"I'm not going to!" said Fenwick.

"But I must!"

Fenwick followed him downstairs, still trying to persuade him of the foolishness of driving back tonight. When Ellerbee heard of it he seemed appalled.

"It's impossible, Dr. Baker! I wouldn't have suggested your not returning if there were any chance of getting through. I assure you there isn't."

"Nevertheless I must try. Dr. Fenwick will remain, and I will come back tomorrow afternoon to complete our investigation. There are important things I must attend to before then, however."

Fenwick had the sudden feeling that Baker was in a flight of panic. His words had an aimless stream-of-consciousness quality that contrasted sharply with his usually precise speech. Fenwick was certain there was nothing sufficiently important that it demanded his attention on a night like this. He could have telephoned his family and had his wife cancel any appointments.

No, Fenwick thought, there was nothing Baker had to go *to*; rather, he was running *from*. He was running in panicky fear from his failure to pin down the hoax in the crystals. He was running, Fenwick thought, from the fear that there might be no hoax.

It seemed incredible that such an experience could trigger so strong a reaction. Yet Fenwick was aware that Baker's attitude toward Ellerbee and his device was not merely one aspect of Baker's character. His attitude in these things *was* his character.

Fenwick dared not challenge Baker with these thoughts. He knew it would be like probing Baker's flesh with a hot wire. There was nothing at all that he could do to stop Baker's flight.

Ellerbee insisted on loaning him a powerful flashlight and a hand lantern, which Baker ridiculed but accepted. It was only after Baker's tail-light had disappeared in the thick mist that Fenwick remembered he still had the crystal cube in his coat pocket.

"He's bound to get stuck and spend the night on the road," said Ellerbee. "He'll be so upset he'll never come back to finish his investigation."

Fenwick suspected this was true. Baker would seal off every association and reminder of the communicator crystals as if they were some infection that would not heal. "There's no use beating your brains out trying to get the NBSD to pay attention," Fenwick told Ellerbee. "You've got a patent. Figure out some gadgety use and start selling the things. You'll get all the attention you want."

"I wanted to do it in a dignified way," said Ellerbee regretfully.

You, too, Fenwick thought as he moved back up the stairs to the spare bedroom.

Fenwick undressed and got into bed. He tried to read a book he had borrowed from Ellerbee, but it held no interest for him. He kept thinking about Baker. What produced a man like Baker? What made him tick, anyway?

Fenwick had practically abandoned his earlier determination that something had to be done about Baker. There was really nothing that could be done about Baker, Bill Baker in particular—and the host of assorted Bakers scattered throughout the world in positions of power and importance, in general.

They stretched on and on, back through the pages of history and time. Jim Ellerbee understood the breed. He had quite rightly tagged Baker in addressing him as "Dear Urban." Pope Urban, who persecuted the great Galileo, had certainly been one of them.

It wasn't that Baker was ignorant or stupid. He was neither. Fenwick gave reluctant respect to his intelligence and his education. Baker was quick-witted. His head was stuffed full of accurate scientific information from diversified fields.

But he refused to be jarred loose from his fixed position that scientific breakthroughs could come from any source but the Established Authority. The possibility that the crackpot fringe could produce such a break-through panicked him. It *had* panicked him. He was fleeing dangerously now through the night, driven by a fear he did not know was in him.

Inflexibility. This seemed to be the characteristic that marked Baker and his kind. Defender of the Fixed Position might well have been his title. With all his might and power, Bill Baker defended the Fixed Position he had chosen, the Fixed Position behind the wall of Established Authority.

A blind spot, perhaps? But it seemed more than mere blindness that kept Baker so hotly defending his Fixed Position. It seemed as if, somehow, he was aware of its vulnerability and was determined to fight off any and all attacks, regardless of consequences.

Fenwick didn't know. He felt as if it was less than hopeless, however, to attempt to change Bill Baker. Any change would have to be brought about by Baker himself. And that, at the moment, seemed far less likely than the well-known snowball in Hades.

Fenwick knew he must have dozed off to sleep with the light still on in the room and Ellerbee's unread book opened over his chest. He did not know what time it was when he awoke. He was aware only of a suffocating sensation as if some ghostly aura were within the room, filling it, pressing down upon him. A wailing of agony and despair seemed to scratch at his senses although he was certain there was no audible sound. And a depression clutched at his soul as if death itself had suddenly walked unseen through the closed door.

Fenwick sat up, shivering in the sudden coolness of the room, but clammy with sweat over his whole body. He had never experienced such sensations before in his life. His stomach turned to a hard ball under the flow of panic that surged through all his nerves.

He forced himself to sit quietly for a moment, trying to release his fear-tightened muscles. He relaxed the panic in his stomach and looked slowly about the room. He could recall no stimulus in his sleep that had produced such a reaction. He hadn't even been dreaming, as far as he could tell. There was no recollection of any sound or movement within the house or outside.

He was calmer after a moment, but that sensation of death close at hand would not go away. He would have been unable to describe it if asked, but it was there. It filled the atmosphere of the room. It seemed to emanate from—

Fenwick turned his head about. It was almost as if there was some definite source from which the ghastly sensation was pouring over him. The walls—the air of the room—

His eyes caught the crystal on the table by the bed.

He could feel the force of death pouring from it.

His first impulse was to pick up the thing and hurl it as far as he could. Then in saner desperation he leaped from the bed and threw on his clothes. He grabbed the crystal in his hand and ran out through the door and down the stairs.

Jim Ellerbee was already there in the living room. He was seated by the old-fashioned library table, his hand outstretched upon it. In his hand lay the counterpart of the crystal Fenwick carried.

"Ellerbee!" Fenwick cried. "What's going on? What in Heaven's name is coming out of these things?"

"Baker," said Ellerbee. "He smashed up on the road somewhere. He's out there dying."

"Can you be sure? Then don't sit there, man! Let's get on our way!"

Ellerbee shook his head. "He'll be dead before we can get there."

"How do you know he cracked up, anyway? Can you read that out of the crystal?"

Ellerbee nodded. "He kept it in his pocket. It's close enough to him to transmit the frantic messages of his dying mind."

"Then we've got to go! No matter if we get there in time or not."

Ellerbee shook his head again. "Sam is on his way over here. He's in touch with Baker. He says he thinks he can talk Baker back."

"*Talk* him back? What do you mean by that?"

Ellerbee hesitated. "I'm not sure. In some ways Sam understands a lot more about these things than I do. He can do things with the crystals that I don't understand. If he says he can talk Dr. Baker back, I think maybe he can."

"But we can't depend on that!" Fenwick said frantically. "Can't we get on our way in the car and let Sam do what he thinks he can while we drive? Maybe he can get Baker to hold on until we get him to a doctor."

"You don't understand," said Ellerbee. "Dr. Baker has gone over the edge. He's *dying*. I know what it's like. I looked into a dying mind once before. There is nothing whatever that a doctor can do after an organism starts dying. It's a definite process. Once started, it's irreversible."

"Then what does Sam—?"

"Sam thinks he knows how to reverse it."

There wasn't much pain. Not as much as he would have supposed. He felt sure there was terrible damage inside. He could feel the warmth of blood welling up inside his throat. But the pain was not there. That was good.

In place of pain, there was a kind of infinite satisfaction and a growing peace. The ultimate magnitude of this peace, which he could sense, was so great that it loomed like some blinding glory.

This was death. The commitment and the decision had been made. But this was better than any alternative. He could not see how there could have been any question about it.

He was lying on his back in the wet clay of a bank below the road. It was raining, softly now, and he rather liked the gentle drop of it on his face. Somewhere below him the hulk of his wrecked car lay on its side. He could smell the unpleasant odor of gasoline. But all of this was less than nothing in importance to him now. Somewhere in the back of his mind was a remnant of memory of what he had been doing this day. He remembered the name of John Fenwick, and the memory brought a faint amusement to his bloody lips. There had been some differences between him and John Fenwick. Those differences were also less than nothing, now. All differences were wiped out. He gave himself up to the pleasure of being borne along on that great current that seemed to be carrying him swiftly to a quiet place.

After a time, he remembered two other names, also. James Ellerbee and Sam Atkins. He remembered a crystal, and it meant nothing. He remembered that it was in his pocket and that for some time he had felt a warmth from it, that was both pleasant and unpleasant. It was of no importance. He might have reached for it and thrown it farther from him, but his arm on that side was broken.

He was glad that there was nothing—nothing whatever—that had any magnitude of importance. Even his family—they were like fragments of a long-ago dream.

He lay waiting quietly and patiently for the swiftly approaching destination of ultimate peace. He did not know how long it would take, but he knew it could not be long, and even the journey was sweet.

It was while he waited, letting his mind drift, that he became aware of the intruder. In that moment, the pain boiled up in shrieking agony.

He had thought himself alone. He wanted above all else to be alone. But there was someone with him. He wasn't sure how he knew. He could simply *feel* the unwanted presence. He strained to see in the wet darkness. He listened for muted sounds. There was nothing. Only the presence.

"Go away!" he whispered hoarsely. "Go away, and leave me alone—whoever you are."

"No. Let me take you by the hand, William Baker. I have come to show you the way back. I have come to lead you back."

"Leave me alone! Whoever you are, leave me alone!" Baker was conscious of his own voice screaming in the black night. And it was not only terror of the unknown presence that made him scream, but the physical pain of crushed bones and torn flesh was sweeping like a torrent through him.

"Don't be afraid of me. You know me. You remember, we met this afternoon. Sam Atkins. You remember, Dr. Baker?"

"I remember." Baker's voice was a painful gasp. "I remember. Now go away and leave me alone. You can do nothing for me. I don't want you to do anything for me."

Sam Atkins. The crystal. Baker wished he could reach the cursed thing and hurl it away from him. That must be how Atkins was communicating with him. Yes, somehow it was possible. He had found no trick, no gimmick. Somehow, the miserable things worked.

But what did Sam Atkins want? He had broken in on a moment that was as private as a dream. There was nothing he could do. Baker was dying. He knew he was dying. There was no medicine that could heal the battering his body had taken. He had been slipping away into peace and release of pain. He had no desire to have it interrupted.

There was no more evidence of Sam Atkins' presence. It was there—and Baker wished furiously that Atkins would let his death be a private thing—but he was not interfering now.

There was the faint suggestion of other presences, too. Baker thought he could pick them out, Fenwick and Ellerbee. They were all gathered to watch him die through the crystals. It was unkind of them to so intrude—but it didn't really matter very much. He began drifting pleasantly again.

"William Baker." The soft voice of Sam Atkins shattered the peaceable realm once more. "We must do some healing before we start back, Dr. Baker. Give me your hand, and come with me, Dr. Baker, while we touch these tissues and heal their breaks. Stay close to me and the pain will not be more than you can endure."

The night remained dark and there was no sound, but Baker's body arched and twisted in panic as he fought against invisible hands that seemed to touch with fleeting, exploratory passes over him.

"I don't want to be healed," he whispered. "There is nothing that can be done. I'm dying. I want to die! Can't you understand that? I want to die! I don't want your help!"

He had said it. And the shock of it jolted even him in the depths of his half-conscious mind. Could a man really *want* to die?

Yes.

He had forgotten what terror he had left so far behind. He knew only that he wanted to move forever in the direction of the flowing peace.

Like probing fingers, Sam Atkins' mind continued to touch him. It scanned the broken organs of his body, and, in some kind of detached way, Baker felt that he was accompanying Atkins on that journey of exploration, even as Sam had asked.

They searched the skeleton and found the splintered bones. They examined the muscle structure and found the torn and shattered tissue. They searched the dark recesses of his vital organs and came to injury that Baker knew was hopeless.

"You built this once," Sam Atkins' voice whispered. "You can build it again. The materials are all here. The blood stream is still moving. The nerve tissue will carry your instructions. I'll supply the scaffolding—while you build—"

He remembered. Baker examined the long-untouched record of when he had done this before. He remembered the construction of cells, the building of organs, the interconnection of nerve tissue. He felt an infinite sadness at the present ruin. Yes—he could build again.

Sam Atkins' face was like that of a dead man. Across the table from him, Jim Ellerbee and John Fenwick watched silently. Faintly, between them was the crystal-projected image of Baker's body.

Fenwick felt the cold touch of some mysterious unknown prickle his scalp. Sam Atkins seemed remote and alien, like the practitioner of ancient and forbidden arts. Fenwick found the question tumbling over and over in his mind, who is this man? He felt as if the very life energy of Sam Atkins was somehow flowing out through the crystal, across space, to the distant broken body of Bill Baker and was supporting it while Baker's own feeble energy was consumed in the rebuilding of his shattered organs.

Though Fenwick and Ellerbee held their own crystals, Sam had somehow shut them out. They were in faint contact with Baker, but they could not follow the fierce contact that Sam's mind held with him.

Ellerbee's face showed worry and a trace of panic. He hesitantly reached out to touch the immobile figure of Sam Atkins, who sat with closed eyes and imperceptible breath. Fenwick sensed disaster. He arrested the motion of Ellerbee's hand.

"I think you could kill them both," he whispered. The life force of one man, divided between two—it was not sufficient to cope with unexpected shocks to either, now.

Ellerbee desisted. "I've never seen anything like this before," he said. "I don't know what Sam's doing—I don't know how he's doing it—"

Fenwick looked sharply at Ellerbee. Ellerbee had discovered the crystals, so he and Sam said. Yet Sam was able to do things with them that Ellerbee could not conceive. Fenwick wondered just who was responsible for the crystals. And he resolved that some day, when and if Baker pulled out of this, he would learn something more about Sam Atkins.

Time moved beyond midnight and into the early morning hours of the day, but this meant nothing to William Baker. He was in the midst of eternity. Because the old pattern was there, and the ancient memories were clear, his reconstruction moved at a pace that was limited only by the materials available. When these grew scarce, Sam Atkins showed him how to break down and utilize other structures that could be rebuilt leisurely at a later time. There was remembered joy in the building and, once started, Baker gave only idle wonder to the question of whether this was more desirable than death. He did not know. This seemed the right thing to do.

In the presence of Sam Atkins everything he was doing seemed right, and a lifetime of doubts, and errors, and fears seemed distant and vague.

But Sam said suddenly, "It is almost finished. Just a little farther and you'll have to go the rest of the way alone."

Terror struck at Baker. He had reached a point where he was absolutely sure he could *not* go on alone without Sam's supporting presence. "You tricked me!" Baker cried. "You tricked me! You didn't tell me I would have to be reborn alone!"

"Doesn't every man?" said Sam. "Is there any way to be born, except alone?"

Slowly, the world closed in about Baker.

Light. Sounds.

Wet. Cold.

The impact of a million idiot minds. The coursing of cosmic-ray particles. The wrenching of Earth's magnetic and gravitational fields. Old and sluggish memories were renewed, memories meant to be buried for all of his life.

Baker felt as if he were suddenly running down a dark and immense corridor. Behind were all the terrors spawned since the beginning of time. Ahead were a thousand openings of light and safety. He raced for the nearest and brightest and most familiar.

"No," said Sam Atkins. "You cannot go that way again. It is the way you went before—and it led to this—to a search for death. For you, it will lead only to the same goal again."

"I can't go on!" Baker cried. The terrors seemed to be swiftly closing in.

"Take my hand a moment longer," said Sam. "Inspect these more distant paths. There are many of them that will be agreeable to you."

Baker felt calmer now in the renewed presence of Sam Atkins. He passed the branching pathway that Sam had forbidden, that had seemed so bright. He sensed now why Sam had cautioned him against it. Far down, in the depths of it, he glimpsed faintly a dark ugliness that he had not seen before. He shuddered.

Directly ahead there seemed to be the opening of a corridor of blazing brightness. Baker's calmness increased as he approached. "This one," he said.

He heard nothing, but he sensed Sam Atkins' smile, and nod of approval.

He remembered now for the first time why he had wanted to die. It was to avoid the very terrors by which he had been pursued through the dark corridor. All this had happened before, and he had gone down the pathway Sam had forbidden. Somehow, like a circle, it had come back to this very point, to this forgotten experience for which he had been willing to die rather than endure again.

It was very bewildering. He did not understand the meaning of it. But he knew he had corrected a former error. He was back in the world. He was alive again.

Sam Atkins looked up at his companions through eyes that seemed all but dead. "He's going to make it," he said. "We can get the car out and pick up Baker now."

They used Sam's panel truck, which had a four-wheel drive and mud tires. Nothing else could possibly get through. Fenwick left his own car at Ellerbee's.

It was still raining lightly as the truck sloshed and slewed through the muck that was hardly recognizable now as a road. For an hour Sam fought the wheel to hold the car approximately in the middle of the brownish ooze that led them through the night. The three men sat in the cab. Behind them, a litter and first-aid equipment had been rigged for Baker. Sam told them nothing would be needed except soap and water, but Fenwick and Ellerbee felt it impossible to go off without some other emergency equipment.

After an hour, Sam said, "He's close. Just around the next bend. That's where his car went off."

Baker loomed suddenly in the lights of the car. He was standing at the edge of the road. He waved an arm wearily.

Fenwick would not have recognized him. And for some seconds after the car had come to a halt, and Baker stood weaving uncertainly in the beam of the lights, Fenwick was not sure it was Baker at all.

He looked like something out of an old Frankenstein movie. His clothes were ripped almost completely away. Those remaining were stained with blood and red clay, and soaked with rain. Baker's face was laced with a network of scars as if he had been slashed with a shower of glass not too long ago and the wounds were freshly healed. Blood was caked and cracked on his face and was matted in his hair.

He smiled grotesquely as he staggered toward the car door. "About time you got here," he said. "A man could catch his death of cold standing out here in this weather."

Dr. William Baker was quite sure he had no need of hospitalization, but he let them settle him in a hospital bed anyway. He had some thinking to do, and he didn't know of a better place to get it done.

There was a good deal of medical speculation about the vast network of very fresh scars on his body, the bones which X rays showed to have been only very recently knit, and the violent internal injuries which gave some evidence of their recent healing. Baker allowed the speculation to go on without offering explanations. He let them tap and measure and apply electrical gadgets to their heart's content. It didn't bother the thinking he had to get done.

Fenwick and Ellerbee came back the next day to see him. The two approached the bed so warily that Baker burst out laughing. "Pull up chairs!" he exclaimed. "Just because you saw me looking a shade less than dead doesn't mean I'm a ghost now. Sit down. And where's Sam? Not that I don't appreciate seeing your ugly faces, but Sam and I have got some things to talk about."

Ellerbee and Fenwick looked at each other as if each expected the other to speak.

"Well, what's the matter?" demanded Baker. "Nothing's happened to Sam, I hope!"

Fenwick spoke finally. "We don't know where Sam is. We don't think we'll be seeing him again."

"Why not?" Baker demanded. But in the back of his mind was the growing suspicion that he knew.

"After your—accident," said Fenwick, "I went back to the farm with Ellerbee and Sam because I'd left my car there. I went back to bed to try to get some more shut-eye, but the storm had started up again and kept me awake. Just before dawn a terrific bolt of lightning seemed to strike Sam's silo. Later, Jim went out to check on his cows and help his man finish up the milking.

"By mid-morning we hadn't heard anything from Sam and decided to go over and talk to him about what we'd seen him do for you. I guess it was eleven by the time we got there."

Jim Ellerbee nodded agreement.

"When we got there," Fenwick went on, "we saw that the front door of the house was open as if the storm had blown it in. We called Sam, but he didn't answer, so we went on in. Things were a mess. We thought it was because of the storm, but then we saw that drawers and shelves seemed to have been opened hastily and cleaned out. Some things had been dropped on the floor, but most of the stuff was just gone.

"It was that way all through the house. Sam's bed hadn't been disturbed. He had either not slept in it, or had gone to the trouble of making it up even though he left the rest of the house in a mess."

"Sounds like the place might have been broken into," said Baker. "Didn't you notify the sheriff?"

"Not after we'd seen what was outside, in back."

"What was that?"

"We wanted to see the silo after the lightning had struck it. Jim said he'd always been curious about that silo. It was one of the best in the county, but Sam never used it. He used a pit.

"When we went out, all the cows were bellowing. They hadn't been milked. Sam did all his own work. Jim called his own man to come and take care of Sam's cows. Then we had a close look at the silo. It had split like a banana peel opening up. It hardly seemed as if a bolt of lightning could have caused it. We climbed over the broken pieces to look inside. It was still warm in there. At least six hours after lightning—or whatever had struck it, the concrete was still warm. The bottom and several feet of the sides of the silo were covered with a glassy glaze."

"No lightning bolt did that."

"We know that now," said Fenwick. "But I had seen the flash of it myself. Then I remembered that in my groggy condition that morning something had seemed wrong about that flash of lightning. Instead of a jagged tree of lightning that formed instantly, it had seemed like a thin thread of light striking *upward*. I thought I must be getting bleary-eyed and tried to forget it. In the silo, I remembered. I told Jim.

"We went back through the house once more. In Sam's bedroom, as if accidently dropped and kicked partway under the bed, I found this. Take a look!"

Fenwick held out a small book. It had covers and pages as did any ordinary book. But when Baker's fingers touched the book, something chilled his backbone.

The material had the feel and appearance of white leather—yet Baker had the insane impression that the cells of that leather still formed a living substance. He opened the pages. Their substance was as foreign as that of the cover. The message—printing, or whatever it might be called—consisted of patterned rows of dots, pin-head size, in color. It reminded him of computer tape cut to some character code. He had the impression that an eye might scan those pages and react as swiftly as a tape-fed computer.

Baker closed the book. "Nothing more?" he asked Fenwick.

"Nothing. We thought maybe you had found out something else when he worked to save your life."

Baker kept his eyes on the ceiling. "I found out a few things," he said. "I could scarcely believe they were true. I have to believe after hearing your story."

"What did you find?"

"Sam Atkins came from—somewhere else. He went back in the ship he had hidden in the silo."

"Where did he come from? What was he doing here?"

"I don't know the name of the world he was from or where it is located. Somewhere in this galaxy, is about all I can deduce from my impressions. He was here on a scientific mission, a sociological study. He was responsible for the crystals. I suppose you know that by now?" Baker glanced at Ellerbee.

Jim Ellerbee nodded. "I suspected for a long time that I was being led, but I couldn't understand it. I thought I was doing the research that produced the crystals, but Sam would drop a hint or a suggestion every once in a while, that would lead off on the right track and produce something fantastic. He knew where we were going, ahead of time. He led me to believe that we were exploring together. Do you know why he did this?"

"Yes," said Baker. "It was part of his project. The project consisted of a study of human reaction to scientific processes which our scientific culture considered impossible. He was interested in measuring our flexibility and reaction to such introductions."

Baker smiled grimly. "We sure gave him his money's worth, didn't we! We really reacted when he brought out his little cubes. I'd like to read the report he writes up!"

"Why did he leave so suddenly?" asked Fenwick. "Was he through?"

"No, that's the bad part of it. My reaction to the crystals was a shock that sent me into a suicidal action—"

Fenwick stared at him, shocked. "You didn't—"

"But I did," said Baker calmly. "All very subconsciously, of course, but I did try to commit suicide. The crystals triggered it. I'll explain how in a minute, but since Sam Atkins was an ethical being he felt the responsibility for what had happened to me. He had to reveal himself to the extent of saving my life—and helping me to change so that the suicidal drive would not appear again. He did this, but it revealed too much of himself and destroyed the chance of completing his program. When he gets back home, he's really going to catch hell for lousing up the works. It's too bad."

Jim Ellerbee let out a long breath. "Sam Atkins—somebody from another world—it doesn't seem possible. What things he could have taught us if he'd stayed!"

Fenwick wondered why it had to have been Baker to receive this knowledge. Baker, the High Priest of the Fixed Position, the ambassador of Established Authority. Why couldn't Sam Atkins—or whatever his real name might be—have whispered just a few words of light to a man willing to listen and profit? His bowels felt sick with the impact of opportunity forever lost.

"How did the crystals trigger a suicidal reaction?" asked Fenwick finally, as if to make conversation more than anything else.

Baker's face seemed to glow. "That's the really important thing I learned from Sam. I learned that about me—about all of us. It's hard to explain. I experienced it—but you can only hear about it."

"We're listening," said Fenwick dully.

"I saw a picture of a lathe in a magazine a few months ago," said Baker slowly. "You can buy one of these lathes for $174,000, if you want one. It's a pretty fancy job. The lathe remembers what it does once, and afterwards can do it again without any instructions.

"The lathe has a magnetic tape memory. The operator cuts the first piece on the lathe, and the tape records all the operations necessary for that production. After that, the operator needs only to insert the metal stock and press the start button.

"There could be a million memories in storage, and the lathe could draw on any one of them to repeat what it had done before at any time in its history."

"I don't see what this has got to do with Sam and you," said Fenwick.

Baker ignored him. "A long time ago a bit of life came into existence. It had no memory, because it was the first. But it faced the universe and made decisions. That's the difference between life and nonlife. Did you know that, Fenwick? The capacity to make decisions without pre-programming. The lathe is not alive because it must be pre-programmed by the operator. We used to say that reproduction was the criterion of life, but the lathe could be pre-programmed to build a duplicate of itself, complete with existing memories, if that were desired, but that would not make it a living thing.

"Spontaneous decision. A single cell can make a simple binary choice. Maybe nothing more complex than to be or not to be. The decision may be conditioned by lethal circumstances that permit only a 'not' decision. Nevertheless, a decision *is* made, and the cell shuts down its life processes in the very instant of death. They are not shut down for it.

"In the beginning, the first bit of life faced the world and made decisions, and memory came into being. The structures of giant protein molecules shifted slightly in those first cells and became a memory of decisions and encounters. The cells split and became new pairs carrying in each part giant patterned molecules of the same structure. These were memory tapes that grew and divided and spread among all life until they carried un-numbered billions of memories.

"Molecular tapes. Genes. The memory of life on earth, since the beginning. Each new piece of life that springs from parent life comes equipped with vast libraries of molecular tapes recording the experiences of life since the beginning.

"Life forms as complex as mammals could not exist without this tape library to draw upon. The bodily mechanisms could not function if they came into existence without the taped memories out of the ages, explaining why each organ was developed and how it should function. Sometimes, part of the tapes *are* missing, and the organism, if it endures, must live without instructions for some function. One human lifetime is too infinitesimally small to relearn procedures that have taken aeons to develop.

"Just as the lathe operator has a choice of tapes which will cause the lathe to function in different ways, so does new life have a choice. The accumulated instructions and wisdom of the whole race may be available, except for those tapes which have been lost or destroyed through the ages. New life has a choice from that vast library of tapes. In its inexperience, it relies on the parentage for the selection of many proven combinations, and so we conclude certain characteristics are 'dominant' or 'inherited,' but we haven't been able to discover the slightest reason why this is so.

"A selection of things other than color of eyes, the height of growth to be attained, the shape of the body must also be made. A choice of modes of facing the exterior world, a choice of stratagems to be used in attaining survival and security in that world, must be made.

"And there is one other important factor: Mammalian life is created in a universe where only life exists. The mammal in the womb does not know of the existence of the external universe. Somewhere, sometime, the first awareness of this external universe arises. In the womb. Outside the womb. Early in fetal life, or late. When and where this awareness comes is an individual matter. But when it comes, it arrives with lethal impact.

"Awareness brings a million sensory invasions—chemical, physical, extrasensory—none of them understood, all of them terrifying.

"This terrible fear that arises in this moment of awareness and non-understanding is almost sufficient to cause a choice of death rather than life at this point. Only because of the developed toughness, acquired through the aeons, does the majority of mammalian life choose to continue.

"In this moment, choices must be made as to how to cope with the external world, how to understand it so as to diminish the fear it inspires. The library of genetic tapes is full of possible solutions. Parental experience is examined, too, and the very sensory impacts that are the source of the terror are inspected to a greater or lesser extent to see how they align with taped information.

"A very basic choice is then made. It may not be a single decision, but, rather, a system of decisions all based on some fundamental underlying principle. And the choice may not be made in an instant. How long a time it may occupy I do not know.

"When the decision has been made, reaction between the individual and the external universe begins and understanding begins to flow into the data storage banks. As data are stored, and successful solutions found in the encounter with the world, fear diminishes. Some kind of equilibrium is eventually reached, in which the organism decides how much fear it is willing to tolerate to venture farther into areas of the unknown, and how much it is willing to limit its experience because of this fear.

"When the decision has been made, and the point of equilibrium chosen, a personality exists. The individual has shaped himself to face the world.

"And nothing short of a Heavenly miracle will ever change that shape!"

"You have said nothing about how the crystal caused you to attempt suicide," said Fenwick.

"The crystal invalidated the molecular tape I had chosen to provide my foundation program for living. The tape was completely shattered, brought to an end. There was nothing left for me to go on."

"Wait a minute!" said Fenwick. "Even supposing this could happen as you describe it, other programs could be selected out of the great number you have described."

"Quite true. But do you know what happens to an adult human being when the program on which his entire life is patterned is destroyed?"

Fenwick shook his head. "What is it like?"

"It's like it was in the beginning, in that moment of first awareness of the external universe. He is aware of the universe, but has no understanding of it. Previous understanding—or what he thought was understanding—has been invalidated, destroyed. The drive to keep living, that was present in that first moment of awareness, has weakened. The strongest impulse is to escape the terror that follows awareness without understanding. Death is the quickest escape.

"This is why men are inflexible. This is why the Urbans cannot endure the Galileos. This is why the Bill Bakers cannot face the Jim Ellerbees. That was what Sam Atkins wanted to find out.

"If a man should decide his basic program is invalid and decide to choose another, he would have to face again the terror of awareness of a world in which understanding does not exist. He would have to return to that moment of first awareness and select a new program in that moment of overwhelming fear. Men are not willing to do this. They prefer a program—a personality—that is defective, that functions with only a fraction of the efficiency it might have. They prefer this to a basic change of programs. Only when a program is rendered absolutely invalid—as mine was by the crystal communicator—is the program abandoned. When that happens, the average man drives his car into a telephone pole or a bridge abutment, or he steps in front of a truck at a street intersection. I drove into a gully in a storm."

"All this would imply that the tape library is loaded with genetic programs that contain basic defects!" said Fenwick.

Baker hesitated. "That's not quite true," he said finally. "The library of molecular tapes does contain a great many false solutions. But they are false not so much because they are defective as because they are obsolete. All of them worked at one time, under some set of circumstances, however briefly. Those times and circumstances may have vanished long since."

"Then why are they chosen? Why aren't they simply passed over?"

"Because the individual organism lacks adequate data for evaluating the available programs. In addition, information may be presented to him which says these obsolete programs are just the ones to use."

Fenwick leaned against the bed and shook his head. "How could a crazy thing like that come about?"

"Cultures become diseased," said Baker. "Sparta was such a one in ancient times. A more psychotic culture has scarcely existed anywhere, yet Sparta prevailed for generations. Ancient Rome is another example. The Age of Chivalry. Each of these cultures was afflicted with a different disease.

"These diseases are epidemic. Individuals are infected before they emerge from the womb. In the Age of Chivalry this cultural disease held out the data that the best life program was based on the concept of Honor. Honor that could be challenged by a mistaken glance, an accidental touch in a crowd. Honor that had to be defended at the expense of life itself.

"Pure insanity. Yet how long did it persist?"

"And our culture?" said Fenwick. "There is such a sickness in our times?"

Baker nodded. "There's a disease in our times. A cultural disease you might call the Great Gray Plague. It is a disease which premises that safety, security, and effectiveness in dealing with the world may be obtained by agreement with the highest existing Authority.

"This premise was valid in the days when disobedience to the Head Man meant getting lost in a bog or eaten by a saber-toothed tiger. Today it is more than obsolete. It is among the most vicious sicknesses that have ever infected any culture."

"And you were sick with it."

"I was sick with it. You remember I said a molecular program is chosen partly on the basis of data presented by parental sources and the spears of invasion from the external world. This data that came to me from both sources said that I could deal with the world by yielding to Authority, by surrounding myself with it as with a shell. It would protect me. I would have stature. My world-problems would be solved if I chose this pattern.

"I chose it well. In our culture there are two areas of Authority, one in government, one in science. I covered myself both ways. I became a Government Science Administrator. You just don't get any more authoritative than that in our day and time!"

"But not everyone employs this as a basic premise!" exclaimed Fenwick.

"No—not everyone, fortunately. In that, may be our salvation. In all times there have been a few infected individuals—Pope Urban, for example. But in his time the culture was throwing off such ills and was surging forward under the impetus of men like Galileo.

"In our own time we are on the other end of the stick. We are just beginning to sink into this plague; it has existed in epidemic form only a few short decades. But look how it has spread! Our civil institutions, always weak to such infection, have almost completely succumbed. Our educational centers are equally sick. Approach them with a new idea and no Ph. D. and see what happens. Remember the Greek elevator engineer who did that a few years ago? He battered his way in by sheer force. It was the only way. He became a nuclear scientist. But for every one of his kind a thousand others are defeated by the Plague."

Fenwick was grinning broadly. He suddenly laughed aloud. "You must be crazy in the head, Bill. You sound just like me!"

Baker smiled faintly. "You are one of the lucky ones. You and Jim. It hasn't hit you. And there are plenty of others like you. But they are defeated by the powerful ones in authority, who have been infected.

"It's less than fifty years since it hit us. It may have five hundred years to run. I think we'll be wiped out by it before then. There must be something that can be done, some way to stamp it out."

"Well," said Fenwick. "You could give Clearwater enough to get us on our feet and running. That would be a start in the right direction."

"An excellent start," said Baker. "The only trouble is you asked for less than half of what you need. As soon as I get back to the office a grant for what you need will be on its way."

William Baker stayed in the hospital two more days. Apart from his family, he asked that no visitors be admitted. He felt as if he were a newborn infant, facing the world with the knowledge of a man—but innocent of experience.

He remembered the days before the accident. He remembered how he dealt with the world in those days. But the methods used then were as impossible to him now as if he were paralyzed. The new methods, found in that bright portal to which Sam Atkins had helped guide him, were untried. He knew they were right. But he had never used them.

He found it difficult to define the postulates he had chosen. The more he struggled to identify them, the more elusive they seemed to become. When he gave up the struggle he found the answer. He had chosen a program that held no fixed postulates. It was based on a decision to face the world as it came.

He was not entirely sure what this meant. The age-old genetic wisdom was still available to guide him. But he was committed to no set path. Fresh decisions would be required at every turn.

A single shot of vaccine could not stem an epidemic. His immunity to the sickness of his culture could not immunize the entire populace. Yet, he felt there was something he could do. He was just not sure what it was.

What could a single man do? In other times, a lone man had been enough to overturn an age. But William Baker did not feel such heroic confidence in his own capacity.

He was not alone, however. There were the John Fenwicks and the Jim Ellerbees who were immune to the great Plague. It was just that William Baker was probably the only man in the world who had ever been infected so completely and then rendered immune. That gave him a look at both sides of the fence, which was an advantage no one else shared.

There was something that stuck in his mind, something that Sam Atkins had said that night when Baker had been reborn. He couldn't understand it. Sam Atkins had said of the molecular program tape that had been broken: When you cease to be fearful of Authority, you become Authority.

The last thing in the whole world William Baker wanted now was to be Authority. But the thought would not leave his mind. Sam Atkins did not say things that had no meaning.

Baker's return to the office of NBSD was an occasion for outpouring of the professional affection which his staff had always tendered him. He knew that there had been a time when this had given him a great deal of satisfaction. He remembered that fiftieth birthday party.

Looking back, it seemed as if all that must have happened to some other man. He felt like a double of himself, taking over positions and prerogatives in which he was a complete impostor.

This was going to be harder than he had anticipated, he thought.

Pehrson especially, it appeared, was going to be difficult. The administrative assistant came into the office almost as soon as Baker was seated at his desk. "It's very good to have you back," said Pehrson. "I think we've managed to keep things running while you've been gone, however. We have rejected approximately one hundred applications during the past week."

Baker grunted. "And how many have you approved?"

"Approval would have had to await your signature, of course."

"O.K., how many are awaiting my signature?"

"It has been impossible to find a single one which had a high enough Index to warrant your consideration."

"I see," said Baker. "So you've taken care of the usual routine without any help from me?"

"Yes," said Pehrson.

"There's one grant left over from before I was absent. We must get that out of the way as quickly as possible."

"I don't recall any that were pending—" said Pehrson in apology.

"Clearwater College. Get me the file, will you?"

Pehrson didn't know for sure whether the chief was joking or not. He looked completely serious. Pehrson felt sick at the sudden thought that the accident may have so injured the chief's mind that he was actually serious.

He sparred. "The Clearwater College file?"

"That's what I said. Bring a set of approval forms, too."

Pehrson managed to get out with a placid mask on his face, but it broke as soon as he reached the safety of his own office. It wasn't possible that Baker was serious! The check that went out that afternoon convinced him it was so.

When Pehrson left the office, Baker got up and sauntered to the window, looking out over the smoke-gray buildings of Washington. The Index, he smiled, remembering it. Five years he and Pehrson had worked on that. It had seemed like quite a monumental achievement when they considered it finished. It had never been really finished, of course. Continuous additions and modifications were being made. But they had been very proud of it.

Baker wondered now, however, if they had not been very shortsighted in their application of the Index. He sensed, stirring in the back of his mind, not fully defined, possibilities that had never appeared to him before.

His speculations were interrupted by Doris. She spoke on the interphone, still in the sweetly sympathetic tone she had adopted for her greetings that morning. Baker suspected this would last at least a full week.

"Dr. Wily is on the phone. He would like to know if you'd mind his coming in this afternoon. Shall I make an appointment or would you rather postpone these interviews for a few days? Dr. Wily would understand, of course."

"Tell him to come on up whenever he's ready," said Baker. "I'm not doing much today."

President George H. Wily, Ph. D., D.Sc., of Great Eastern University. Wily was one of his best customers.

Baker guessed that he had given Wily somewhere around twelve or thirteen million dollars over the past decade. He didn't know exactly what Wily had done with all of it, but one didn't question Great Eastern's use of its funds. Certainly only the most benevolent use would be made of the money.

Baker reflected on his associations with Wily. His satisfaction had been unmeasurable in those exquisite moments when he had had the pleasure of handing Wily a check for two or three million dollars at a time. In turn, Wily had invited him to the great, commemorative banquets of Great Eastern. He had presented Baker to the Alumni and extolled the magnificent work Baker was doing in the advancement of the cause of Science. It had been a very pleasant association for both of them.

The door opened and Doris ushered Wily into the room. He came forward with outstretched hands. "My dear Baker! Your secretary said you had no objection to my coming up immediately, so I took advantage of it. I didn't hear about your terrible accident until yesterday. It's so good to know that you were not more seriously hurt."

"Thanks," said Baker. "It wasn't very bad. Come and sit down."

Wily was a rather large, beetle-shaped man. He affected a small, graying beard that sometimes had tobacco ashes in it.

"Terrible loss to the cause of Science if your accident had been more serious," Wily was saying. "I don't know of anyone who occupies a more critical position in our nation's scientific advance than you do."

This was what had made him feel safe, secure, able to cope with the problems of the world, Baker reflected. Wily represented Authority, the highest possible Authority in the existing scientific culture.

But it had worked both ways, too. Baker had supplied a similar counterpart for Wily. His degrees matched Wily's own. He represented both Science and Government. The gift of a million dollars expressed confidence on the part of the Government that Wily was on the right track, that his activity was approved.

A sort of mutual admiration society, Baker thought.

"I suppose you are interested in the progress on your application for renewal of Great Eastern's grants," said Baker.

Wily waved the subject away with an emphatic gesture. "Not business today! I simply dropped in for a friendly chat after learning of your accident. Of course, if there is something to report, I wouldn't mind hearing it. I presume, however, the processing is following the usual routine."

"Not quite," said Baker slowly. "An increasing flood of applications is coming in, and I'm finding it necessary to adopt new processing methods to cope with the problem."

"I can understand that," said Wily. "And one of the things I have always admired most about your office is your ability to prevent wastage of funds by nonqualified people. Qualifications in the scientific world are becoming tighter every day. You have no idea how difficult it is to get people with adequate backgrounds today. Men of stature and authority seem to be getting rarer all the time. At any rate, I'm sure we are agreed that only the intellectual elite must be given access to these funds of your Bureau, which are limited at best."

Baker continued to regard Wily across the desk for a long moment. Wily was one of them, he thought. One of the most heavily infected of all. Surround yourself with Authority. Fold it about you like a shell. Never step beyond the boundaries set by Authority. This was George H. Wily, President of Great Eastern University. This was a man stricken by the Great Gray Plague.

"I need a report," said Baker. "For our new program of screening I need a report of past performance under our grants. The last two years would be sufficient, I think, from Great Eastern."

Wily was disturbed. He frowned and hesitated. "I'm sure we could supply such a report," he said finally. "There's never been any question—"

"No question at all," said Baker. "I just need to tally up the achievements made under recent grants. I shall also require some new information for the Index. I'll send forms as soon as they're ready."

"We'll be more than glad to co-operate," said Wily. "It's just that concrete achievement in a research program is sometimes hard to pin point, you know. So many intangibles."

"I know," said Baker.

When Wily was gone, Baker continued sitting at his desk for a long time. He wished fervently that he could talk with Sam Atkins for just five minutes now. And he hoped Sam hadn't gotten too blistered by his mentors when he returned home after fluffing the inquiry he was sent out on.

There was no chance, of course, that Baker would ever be able to talk with Sam again. That one fortuitous encounter would have to do for a lifetime. But Sam's great cryptic statement was slowly beginning to make sense: When you cease to be fearful of Authority, you become Authority.

Neither Baker or Wily, or any of the members of Wily's lock-step staff were Authority. Rather, they all gave obeisance to the intangible Authority of Science, and stood together as self-appointed vicars of that Authority, demanding penance for the slightest blasphemy against it. And each one stood in living terror of such censure.

The same ghost haunted the halls of Government. The smallest civil servant, in his meanest incivility, could invoke the same reverence for that unseen mantle of Authority that rested, however falsely, on his thin shoulders.

The ghost existed in but one place, the minds of the victims of the Plague. William Baker had ceased to recognize or give obeisance to it. He was beginning to understand the meaning of Sam Atkins' words.

He was quite sure the grants to Great Eastern were going to diminish severely.

Within six months, the output from Clearwater College was phenomenal. The only string that Baker had attached to his grants was the provision that the National Bureau of Scientific Development be granted the privilege of announcing all new inventions, discoveries, and significant reports. This worked to the advantage of both parties. It gave the college the prestige of association in the press with the powerful Government agency, and it gave Baker the association with a prominent scientific discovery.

During the first month of operation under the grant, Fenwick appointed a half dozen "uneducated" professors to his physical science staff. These were located with Baker's help because they had previously applied to NBSD for assistance.

The announcement of the developments of the projects of these men was a kind of unearned windfall for both Baker and Fenwick because most of the work had already been done in garages and basements. But no one objected that it gave both Clearwater and NBSD a substantial boost in the public consciousness.

During this period, Baker found three other small colleges of almost equal caliber with Clearwater. He made substantial grants to all of them and watched their staffs grow in number and quality of background that would have shocked George Wily into apoplexy. Baker's announcements of substantial scientific gains became the subject of weekly press conferences.

And also, during this time, he lowered the ax on Great Eastern and two other giants whose applications were pending. He cut them to twenty per cent of what they were asking. A dozen of the largest industrial firms were accorded similar treatment.

Through all this, Pehrson moved like a man in a nightmare. His first impulse had been to resign. His second was to report the gross mismanagement of NBSD to some appropriate congressman. Before he did either of these things the reports began to come in from Clearwater and other obscure points.

Pehrson was a man in whom allegiance was easily swayed. His loyalty was only for the top man of any hierarchy, and he suddenly began to regard Baker with an amazed incredulity. It seemed akin to witchcraft to be able to pull out works of near genius from the dross material Baker had been supporting with his grants. Pehrson wasn't quite sure how it had been done although he had been present throughout the whole process. He only knew that Baker had developed a kind of prescience that was nothing short of miraculous, and from now on he was strictly a Baker man.

Baker was happy with this outcome. The problem of Pehrson had been a bothersome one. Civil Service regulations forbade his displacement. Baker had been undecided how to deal with him. With Pehrson's acceptance of the new methods, the entire staff swung behind Baker, and the previous grumblings and complaints finally ceased. He stood on top in his own office, at least, Baker reflected.

George H. Wily was not happy, however. He waited two full days after receiving the announcement of NBSD's grant for the coming year. He consulted with his Board of Regents and then took a night plane down to Washington to see Baker.

He was coldly formal as he entered Baker's office. Baker shook his hand warmly and invited him to sit down.

"I was hoping you'd drop in again when you came to town," said Baker. "I was sorry we had to ask you for so much new information, but I appreciate your prompt response."

Wily's eyes were frosty. "Is that why you gave us only two hundred thousand?" he asked.

Baker spread his hands. "I explained when you were here last that we were getting a flood of applications. We have been forced to distribute the money much more broadly than in other years. There is only so much to go around, you know."

"There is just as much as you've ever had," snapped Wily. "I've checked on your overall appropriation. And there is no increase in qualified applicants. There is a decrease, if anything.

"I've done a little checking on the grants you've made, Baker. I'd like to see you defend your appropriation for that miserable little school called Clearwater College. I made a detailed study of their staff. They haven't a single qualified man. Not one with a background any better than that of your elevator operator!"

Baker looked up at the ceiling. "I remember an elevator man who became quite a first rate scientist."

Wily glared, waiting for explanation, then snorted. "Oh, *him*—"

"Yes, *him*," said Baker.

"That doesn't explain your wasting of Government funds on such an institution as Clearwater. It doesn't explain your grants to—"

"Let me show you what does explain my grants," said Baker. "I have what I call the Index—with a capital I, you know—"

"I don't care anything about your explanations or your Index!" Wily exclaimed. "I'm here to serve notice that I represent the nation's interest as well as that of Great Eastern. And I am not going to stand by silently while you mismanage these sacred funds the way you have chosen to do in recent months. I don't know what's happened to you, Baker. You were never guilty of such mistakes before. But unless you can assure me that the full normal grant can be restored to Great Eastern, I'm going to see that your office is turned inside out by the Senate Committee on Scientific Development, and that you, personally, are thrown out."

Wily glared and breathed heavily after his speech. He sat waiting for Baker's answer.

Baker gave it when Wily had stopped panting and turned to drumming his fingers on the desk. "Unless your record of achievement is better this year than it has been in recent years, Great Eastern may not get any allotment at all next year," he said quietly.

Wily shaded toward deep red, verging on purple, as he rose. "You'll regret this, Baker! This office belongs to American Science. I refuse to see it desecrated by your gross mismanagement! Good day!"

Baker smiled grimly as Wily stormed out. Then he picked up the phone and asked Doris to get Fenwick at Clearwater. When Fenwick finally came on, Baker said, "Wily was just here. I expected he would be the one. This is going to be it. Send me everything you've got for release. We're going to find out how right Sam Atkins was!"

He called the other maverick schools he'd given grants, and the penny ante commercial organizations he'd set on their feet. He gave them the same message.

It wasn't going to be easy or pleasant, he reflected. The biggest guns of Scientific Authority would be trained on him before this was over.

Drew Pearson had the word even before it reached Baker. Baker read it at breakfast a week after Wily's visit. The columnist said, "The next big spending agency to come under the fire of Congressional Investigation is none other than the high-echelon National Bureau of Scientific Development. Dr. William Baker, head of the Agency, has been accused of indiscriminate spending policies wholly unrelated to the national interest. The accusers are a group of elite universities and top manufacturing organizations that have benefited greatly from Baker's handouts in years past. This year, Baker is accused of giving upwards of five million dollars to crackpot groups and individuals who have no standing in the scientific community whatever.

"If these charges are true, it is difficult to see what Dr. Baker is up to. For many years he has had an enviable record as a tight-fisted, hard-headed administrator of these important funds. Congress intends to find out what's going on. The watchdog committee of Senator Landrus is expected to call an investigation early next week."

Baker was notified that same afternoon.

Senator Landrus was a big, florid man, who moved about a committee hearing chamber with the ponderous smoothness of a luxury liner. He was never visited by a single doubt about the rightness of his chosen course—no matter how erratic it might appear to an onlooker. His faith in his established legislative procedures and in the established tenets of Science was complete.

Since he wore the shield of both camps, his confidence in the path of Senator Robert Landrus was also unmarred by questions.

Baker had faced him many times, but always as an ally. Now, recognizing him as the enemy, Baker felt some small qualms, not because he feared Landrus, but because so much was at stake in this hearing. So much depended on his ability to guide the whims and uncertainties of this mammoth vessel of Authority.

There was an unusual amount of press interest in what might have seemed a routine and unspectacular hearing. No one could recall a previous occasion when the recipients had challenged a Government handout agency regarding the size of the handouts. While Landrus made his opening statement several of the reporters fiddled with the idea of a headline that said something about biting the hand that feeds. It wouldn't quite come off.

Wily was invited to make his statement next, which he did with icy reserve, never once looking in Baker's direction. He was followed by two other university presidents and a string of laboratory directors. The essence of their remarks was that Russia was going to beat the pants off American researchers, and it was all Baker's fault.

This recital took up all of the morning and half the afternoon of the first day. A dozen or so corporation executives were next on the docket with complaints that their vast facilities were being hamstrung by Baker's sudden switch of R & D funds to less qualified agents. Baker observed that the ones complaining were some of those who had never spent a nickel on genuine research until the Government began buying it. He knew that Landrus had not observed this fact. It would have to be called to the senator's attention.

By the end of the day, Landrus looked grave. It was obvious that he could see nothing but villainy in Baker's recent performance. It had been explained to him in careful detail by some of the most powerful men in the nation. Baker was certainly guilty of criminal negligence, if not more, in derailing these funds which Congress had intended should go to the support of the nation's scientific leaders. Landrus felt a weary depression. He hadn't really believed it would turn out this bad for Baker, for whom he had had a considerable regard in times past.

"You have heard the testimony of these witnesses," Landrus said to Baker. "Do you wish to reply or make a statement of your own, Dr. Baker?"

"I most certainly do!" said Baker.

Landrus didn't see what was left for Baker to say. "Testimony will resume tomorrow at nine a.m.," he said. "Dr. Baker will present his statement at that time."

The press thought it looked bad for Baker, too. Some papers accused him openly of attempting to sabotage the nation's research program. Wily and his fellows, and Landrus, were commended for catching this defection before it progressed any further.

Baker was well aware he was in a tight spot, and one which he had deliberately created. But as far as he could see, it was the only chance of utilizing the gift that Sam Atkins had left him. He felt confident he had a fighting chance.

His battery of supporters had not even been noticed in the glare of Wily's brilliant assembly, but Fenwick was there, and Ellerbee. Fenwick's fair-haired boy, George, and a half dozen of his new recruits were there. Also present were the heads of the other maverick schools like Clearwater, and the presidents—some of whom doubled as janitors—of the minor corporations Baker had sponsored.

Baker took the stand the following morning, armed with his charts and displays. He looked completely confident as he addressed Landrus and the assembly.

"Gentlemen—and ladies—" he said. "The corner grocery store was one of America's most familiar and best loved institutions a generation or two ago. In spite of this, it went out of business because we refused to support it. May I ask why we refused to continue to support the corner grocery?

"The answer is obvious. We began to find better bargains elsewhere, in the supermarket. As much as we regret the passing of the oldtime grocer I'm sure that none of us would seriously suggest we bring him back.

"For the same reason I suggest that the time may have come to reconsider the bargains we have been getting in scientific developments and inventions. Americans have always taken pride in driving a good, hard, fair bargain. I see no reason why we should not do the same when we go into the open market to buy ideas.

"Some months ago I began giving fresh consideration to the product we were buying with the millions of dollars in grants made by NBSD. It was obvious that we were buying an impressive collection of shiny, glass and metal laboratories. We were buying giant pieces of laboratory equipment and monstrous machines of other kinds. We were getting endless quantities of fat reports—they fill thousands of miles of microfilm.

"Then I discovered an old picture of what I am sure all unbiased scientists will recognize as the world's greatest laboratory—greatest in terms of measurable output. I brought this picture with me."

Baker unrolled the first of his exhibits, a large photographic blowup. The single, whitehaired figure seated at a desk was instantly recognized. Wily and his group glanced at the picture and glared at Baker.

"You recognize Dr. Einstein, of course," said Baker. "This is a photograph of him at work in his laboratory at the Institute for Advanced Study at Princeton."

"We are all familiar with the appearance of the great Dr. Einstein," said Landrus. "But you are not showing us anything of his laboratory, as you claimed."

"Ah, but I am!" said Baker. "This is all the laboratory Dr. Einstein ever had. A desk, a chair, some writing paper. You will note that even the bookshelves behind him are bare except for a can of tobacco. The greatest laboratory in the world, a place for a man's mind to work in peace. Nuclear science began here."

Wily jumped to his feet. "This is absurd! No one denies the greatness of Dr. Einstein's work, but where would he have been without billions of dollars spent at Oak Ridge, Hanford, Los Alamos, and other great laboratories. To say that Dr. Einstein did not use laboratory facilities does not imply that vast expenditures for laboratories are not necessary!"

"I should like to reverse your question, Dr. Wily, and then let it rest," said Baker. "What would Oak Ridge, Hanford, and Los Alamos have done without Dr. Einstein?"

Senator Landrus floated up from his chair and raised his hands. "Let us be orderly, gentlemen. Dr. Baker has the floor. I should not like to have him interrupted again, please."

Baker nodded his thanks to the senator. "It has been charged," Baker continued, "that the methods of NBSD in granting funds for research have changed in recent times. This is entirely correct, and I should first like to show the results of this change."

He unrolled a chart and pinned it to the board behind him. "This chart shows what we have been paying and what we have been getting. The black line on the upper half of the chart shows the number of millions of dollars spent during the past five years. Our budget has had a moderately steady rise. The green line shows the value of laboratories constructed and equipment purchased. The red line shows the measure of new concepts developed by the scientists in these laboratories, the improvement on old concepts, and the invention of devices that are fundamentally new in purpose or function."

The gallery leaned forward to stare at the chart. From press row came the popping of flash cameras. Then a surge of spontaneous comment rolled through the chamber as the audience observed the sharp rise of the red line during the last six months, and the dropping of the green line.

Wily was on his feet again. "An imbecile should be able to see that the trend of the red line is the direct result of the previous satisfactory expenditures for facilities. One follows the other!"

Landrus banged for order.

"That's a very interesting point," said Baker. "I have another chart here"—he unrolled and pinned it—"that shows the output in terms of concepts and inventions, plotted against the size of the grants given to the institution."

The curve went almost straight downhill.

Wily was screaming. "Such data are absolutely meaningless! Who can say what constitutes a new idea, a new invention? The months of groundwork—"

"It will be necessary to remove any further demonstrators from the hearing room," said Landrus. "This will be an orderly hearing if I have to evict everyone but Dr. Baker and myself. Please continue, doctor."

"I am quite willing for my figures and premises to be examined in all detail," said Baker. "I will be glad to supply the necessary information to anyone who desires it at the close of this session. In the meantime, I should like to present a picture of the means which we have devised to determine whether a grant should be made to any given applicant.

"I am sure you will agree, Senator Landrus and Committee members, that it would be criminal to make such choices on any but the most scientific basis. For this reason, we have chosen to eliminate all elements of bias, chance, or outright error. We have developed a highly advanced scientific tool which we know simply as The Index."

Baker posted another long chart on the wall, speaking as he went. "This chart represents the index of an institution which shall remain anonymous as Sample A. However, I would direct Dr. Wily's close attention to this exhibit. The black median line indicates the boundary of characteristics which have been determined as acceptable or nonacceptable for grants. The colored areas on either side of the median line show strength of the various factors represented in any one institution. The Index is very simple. All that is required is that fifty per cent of the area above the line be colored in order to be eligible for a grant. You will note that in the case of Sample A the requirement is not met."

Fenwick couldn't believe his eyes. The chart was almost like the first one he had ever seen, the one prepared for Clearwater College months ago. He hadn't even known that Baker was still using the idiotic Index. Something was wrong, he told himself—all wrong.

"The Index is a composite," Baker was saying; "the final resultant of many individual charts, and it is the individual charts that will show you the factors which are measured. These factors are determined by an analysis of information supplied directly by the institution.

"The first of these factors is admissions. For a college, it is admission as a student. For a corporation, it is admission as an employee. In each case we present the qualifications of the following at college age: Thomas Edison, Michael Faraday, Nicholai Tesla, James Watt, Heinrich Hertz, Kepler, Copernicus, Galileo, and Henry Ford. The admissibility of this group of the world's scientific and the inventive leaders is shown here." Baker pointed to a minute dab of red on the chart.

"Gentlemen of the Committee," he said, "would you advise me to support with a million-dollar grant an institution that would close its doors to minds like those of Edison and Faraday?"

The roar of surf seemed to fill the committee room as Landrus banged in vain on the table. Photographers' flashes lit the scene with spurts of lightning. Wily was on his feet screaming, and Baker thought he heard the word, "Fraud!" repeated numerous times. Landrus was finally heard, "The room will be cleared at the next outburst!"

Baker wondered if he ever did carry out such a threat.

But Wily prevailed. "No such question was ever asked," he cried. "My organization was never asked the ridiculous question of whether or not it would admit these men. Of course we would admit them if they were known to us!"

"I should like to answer the gentleman's objection," Baker said to Landrus.

The senator nodded reluctantly.

"We did not, of course, present these men by name. That would have been too obvious. We presented them in terms of their qualifications at the age of college entrance. You see how many would have been turned down. How many, therefore, who are the intellectual equals of these men are also being turned down? Dr. Wily says they would be admitted if they were known. But of course they could not be known at the start of their careers!"

Baker turned the chart and quickly substituted another. "The second standard is that of creativeness. We simply asked the applicants to describe ten or more new ideas of speculations entertained by each member of the staff during the past year. When we received this information, we did not even read the descriptions; we merely plotted the degree of response. As you see, the institution represented by Sample A does not consider itself long on speculative ideas."

A titter rippled through the audience. Baker saw Wily poised, beet-red, to spring up once more; then apparently he thought better of it and slumped in his seat.

"Is this a fair test?" Baker asked rhetorically. "I submit that it is. An institution that is in the business of fostering creativeness ought to be guilty of a few new ideas once in a while!"

He changed charts once more and faced the listeners. "We have more than twenty such factors that go into the composition of the Index. I will not weary you with a recital of all of them, but I will present just one more. We call this the area of communication, and it is plotted here for Sample institution A."

Again, a dismal red smudge showed up at the bottom of the sheet. Fenwick could hardly keep from chuckling aloud as he recalled the first time he had seen such a chart. He hoped Baker was putting it over. If the reaction of the gallery were any indication, he was doing so.

"A major activity of scientists in all ages has been writing reports of their activities. If a man creates something new and talks only to himself about it, the value of the man and his discovery to the world is a big round zero. If a man creates something new and tells the whole world about it, the value is at a maximum. Somewhere in between these extremes lies the communicative activity of the modern scientist.

"There was a time when the scientist was the most literate of men, and the writing of a scientific report was a work of literary art. The lectures of Michael Faraday, Darwin's account of his great research—these are literate reading still.

"There are few such men among us today. The modern scientists seldom speak to you and me, but only to each other. To the extent their circle of communication is limited, so is their value. Shall we support the man who speaks to the world, or the man who speaks only in order to hear his own echo?"

He had them now, Fenwick was convinced. He could quit any time and be ahead. The gallery was smiling approval. The press was nodding and whispering to each other. The senators wouldn't be human if they weren't moved.

Baker swept aside all these charts now and placed another series before the audience. "This is the Index on an institution to whom we have given a sizable grant," he said. "Is there anyone here who would question our decision?

"This institution would have accepted every one of the list of scientists I gave you a moment ago. They would have had their chance here. This institution has men in whom new ideas pop up like cherry blossoms in the spring. I don't know how many of them are good ideas. No one can tell at this stage, but, at least, these men are *thinking*—which is a basic requirement for producing scientific discovery.

"Finally, this institution is staffed by men who can't be shut up. They don't communicate merely with each other. They talk about their ideas to anyone who comes along. They write articles for little publications and for big ones. They are in the home mechanics' journals and on publishers' book lists.

"Most important of all, these are some of the men responsible for the red line on the first curve I showed you. These are the men who have produced the most new developments and inventions with the least amount of money.

"I leave it to you, gentlemen. Has the National Bureau of Scientific Development chosen correctly, or should we return to our former course?"

There were cheers and applause as Baker sat down. Landrus closed the hearing with the announcement that the evidence would be examined at length and a report issued. Wily hurried forward to buttonhole him as the crowd filed out.

"It was a good show," Fenwick said, "but I'm still puzzled by what you've done. This new Index is really just about as phony as your old one."

They were seated in Baker's office once more. Baker smiled and glanced through the window beyond Fenwick. "I suppose so," Baker admitted finally, "but do you think Wily will be able to convince Landrus and his committee of that no matter how big a dinner he buys him tonight?"

"No—I don't think he will."

"Then we've accomplished our purpose. Besides, there's a good deal of truth buried in the Index. It's no lie that we can give them scientific research at a cheaper price than ever before."

"But what was the purpose you were trying to accomplish?"

Baker hesitated. "To establish myself as an Authority," he said, finally. "After today, I will be the recognized Authority on how to manage the nation's greatest research and development program."

Fenwick stared, then gasped. "Authority—you? This is the thing you were trying to fight. This is the great Plague Sam Atkins taught you—"

Baker was shaking his head and laughing. "No. Sam Atkins didn't tell me that one man could become immune and fight the Plague head on all by himself. He taught me something else that I didn't understand for a long time. He told me that he who ceases to fear Authority becomes Authority.

"To become Authority was the last thing in the world I wanted. But finally I recognized what Sam meant; it was the only way I could ever accomplish anything in the face of this Plague. You can't tell men of this culture that it is wrong to put themselves in total agreement with Authority. If that's the program on which they've chosen to function, the destruction of the program would destroy them, just as it did me. There had to be another way.

"If men are afraid of lions, you don't teach them it's wrong for men to be afraid of beasts; you teach them how to trap lions.

"If men are afraid of new knowledge-experiences, you don't teach them that new knowledge is not to be feared. There was a time when men got burned at the stake for such efforts. The response today is not entirely different. No—when men are afraid of knowledge you teach them to trap knowledge, just as you might teach them to trap lions.

"I can do this now because I have shown them that I am an Authority. I can lead them and it will not fracture their basic program tapes, which instruct them to be in accord with Authority. I can stop their battle against those who are not possessed of the Plague. It may even be that I can change the course of the Plague. Who knows?"

Fenwick was silent for a long time. Then he spoke again. "I read somewhere about a caterpillar that's called the Processionary Caterpillar. Several of them hook up, nose to fanny, and travel through a forest wherever the whims of the front caterpillar take them.

"A naturalist once took a train of Processionary Caterpillars and placed them on the rim of a flower pot in a continuous chain. They marched for days around the flower pot, each one supposing the caterpillar in front of him knew where he was going. Each was the Authority to the one behind. Food and water were placed nearby, but the caterpillars continued marching until they dropped off from exhaustion."

Baker frowned. "And what's that got to do with—?"

"You," said Fenwick. "You just led the way down off the flower pot. You just got promoted to head caterpillar."